HANDBOOK OF SERIAL COMMUNICATIONS INTERFACES

HANDBOOK OF SERIAL COMMUNICATIONS INTERFACES

A COMPREHENSIVE COMPENDIUM OF SERIAL DIGITAL INPUT/OUTPUT (I/O) STANDARDS

LOUIS E. FRENZEL Jr

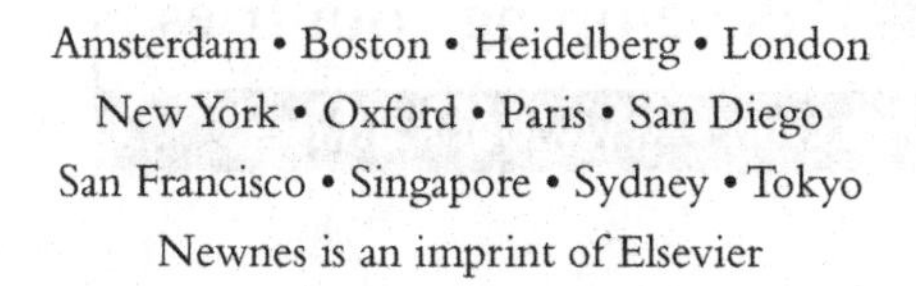

Newnes is an imprint of Elsevier
The Boulevard, Langford Lane, Kidlington, Oxford OX5 1GB, UK
225 Wyman Street, Waltham, MA 02451, USA

ISBN: 978-0-12-800629-0

British Library Cataloguing-in-Publication Data
A catalogue record for this book is available from the British Library.

Library of Congress Cataloging-in-Publication Data
A catalog record for this book is available from the Library of Congress.

For Information on all Newnes publications
visit our website at http://store.elsevier.com/

Printed and bound in the United States

CONTENTS

Section II Medium-Speed Interfaces (10 Mb/s to 1 Gb/s)

Section III High-Speed Interfaces (1–100 Gb/s)

Section IV Broadband Interfaces

Section V Wireless Interfaces

PREFACE

Working with microcontrollers or any other electronic equipment involves working with interfaces, the electronic circuits that connect one device to another for communications. If you are designing electronic equipment, you have probably already worked with one or more interfaces. The most common wired interfaces, for example, are USB and HDMI in consumer equipment and RS-232/RS-485 in industrial equipment. However, there are dozens of others, some of which may not be familiar. Over the years, many new serial interfaces have been developed for specific purposes. This book attempts to catalog all of the more common well-known interfaces, both wired and wireless. This compendium of interfaces will serve as a guide for selecting, comparing, and using serial interfaces.

Several of the chapters are designed to provide a concise summary of related data communications and networking techniques. The chapters cover wired baseband interfaces, broadband fundamentals, and wireless theory and techniques. You probably already have the necessary technical background to use this book, but if not these chapters will give you a short course on the basics. The remainder of the chapters are devoted to the individual interface and protocol descriptions. A standard format is used for each to facilitate comparisons.

Please note that the interface descriptions are short and to the point with the key data that you need to know. Details of each interface or related protocols are not included. The interface summaries help you focus on the main specifications and applications for comparison purposes. If you need more details, go to the original sources of the standards or available books.

The book is organized into focused sections. The first three sections cover wired baseband interfaces listing the interfaces by their data rate: slow, medium, and high. To use these chapters, start by estimating the speed need of your application and then go to the related section to look through the available interfaces as you focus on the application. The interfaces are alphabetized per section. Each interface is a short chapter. Then select the best choice for your project.

Another section focuses on broadband interfaces that use modulation techniques. A review of the most commonly used modulation methods is included. Finally there is a chapter devoted to wireless basics that explains the fundamentals needed to use the wireless interface chapters.

One important point about the interface listings is that only the physical and data link layers of the OSI model (see Chapter 2) are used in describing the interface. In some cases, the related higher layer protocols may be briefly described. Sometimes a common physical interface is used with several different protocols. These protocol variations are briefly mentioned but not fully described due to their complexity.

This book is not intended to cover each interface standard in detail. Most interface specifications are dozens or even hundreds of pages long and impossible to cover all of them in a single book. Many interfaces are proprietary or require membership in an organization to access. As a result, this book offers a summary of the most common interfaces that lets you discover the one most applicable to your needs and compare similar interfaces. I probably missed some of the available interfaces. If you know of any that need to be included, let me know and I will cover them in a future edition.

I hope you will find the book useful. I am open to feedback for any corrections you may want to submit or for any interfaces or features that should be added.

Best wishes,
Louis E. Frenzel Jr
Austin, TX
2015

CHAPTER ONE

Introduction to Serial I/O Communications

The very first electrical/electronic system ever created was the telegraph invented by Samuel Morse in 1844. It predates electric light bulbs, the telephone, and radio. And it was a serial input/output (I/O) system using the dots and dashes of Morse code. Radio followed in the late nineteenth and early twentieth century with Marconi's telegraph system also using Morse code. Teletype machines came next in the early twentieth century using the 5-bit Bardot code and typewriter-like electro-mechanical terminals. Wireless telegraphy was widely used for marine and worldwide telegraph services.

It wasn't until the early 1960s when the first real binary serial interfaces were developed and deployed. An example is the RS-232 standard that was developed and used as the interface for teletypewriter terminals and the earliest CRT terminals on mainframe computers. After that many different serial interfaces were developed for specific purposes. Today there are literally dozens of different serial I/O interfaces for virtually any application from low-speed industrial equipment to super high-speed fiber optics systems.

RATIONALE

Digital data is transferred from one point to another in two ways, parallel and serial. In a parallel transfer, all bits move at once from source to destination. In a serial transfer, the bits are sent one at a time. That makes a parallel transfer faster but also requires multiple lanes or paths, one per bit. Parallel transfers are more expensive as they require more hardware. If wires are used, a large complex cable or bus is needed.

Serial data transfers require only a single path or cable so less circuitry is needed. Since only a single path is required, wireless is an option. The cost is less with a serial connection and today speed is not usually a limiting factor.

Handbook of Serial Communications Interfaces.
Doi: http://dx.doi.org/10.1016/B978-0-12-800629-0.00001-2

As digital circuits speeds have increased over the years with smaller faster ICs and processors, parallel transfers over multiple lane buses have reached the limits of their capability. Stray and distributed inductance and capacitance plus crosstalk on cables and printed circuit board bus conductors have become so large that the higher transfer rates needed cannot be achieved. Timing skew from line to line can cause errors.

On the other hand, the faster circuits have made serial transfers more practical. Circuitry and cables are simpler and less expensive. And to use wireless or fiber optic cable, serial data is mandatory.

The outcome of these conditions has been the development of an extensive set of serial I/O methods and standards. The interfaces using these standards are deployed in almost all electronic equipment today. These serial interfaces connect equipment to other equipment, equipment to networks, printed circuit boards (PCBs) to one another, and one IC chip to another.

Because there are so many serial interfaces, engineers are often not aware of all the available alternatives. Furthermore, some interfaces are competitive with others and the question becomes, which is best for my application? Some interfaces are for specialized use while others are very versatile. In addition, technicians or experimenters working on equipment may not know the details of the interfaces they are working with thereby hampering troubleshooting or testing efforts.

The solution is this book that catalogs all the most popular and commonly used interfaces and provides a summary of the specifications and the standards. Engineers and techs can use the book to help select an interface for a new design, verify that an interface is working correctly or just familiarize themselves with a new interface.

APPROACH

The second chapter in this book summarizes the fundamentals of serial data communications. It covers data speed, synchronous versus asynchronous, line coding, error correction, physical media, and the basic circuitry involved.

Next, the most popular serial interfaces are categorized into one of three groups: low speed (0 to 10 Mb/s), medium speed (10 to 1 G/s) or high speed (1 Gb/s to 100 Gb/s). Each interface is then presented in a standard format giving trade name, standard, speed, coding, medium, protocol, common applications, and other relevant data.

Sufficient data is provided on each interface so that the best applications and most likely candidates can be identified. Most protocols are summarized because of the complexities of the standards. However, sufficient details are given so that the specifications can be compared to enable a smart selection.

Later chapters cover wired broadband interfaces that use modulation as well as the most common wireless interfaces.

CHAPTER TWO

Serial I/O Primer: A Short Course in Data Communications and Networking

If you are reading this book, then you probably already know the basics of serial I/O. However, if you want a quick review, this chapter will refresh your memory on the key concepts. The chapter covers the principles of serial data transfer common to all the interfaces to be discussed later. These basics include serial versus parallel transfer, balanced versus unbalanced connections, physical media types, data rate, line coding, protocols, serdes, and other common characteristics. The chapter is not intended to cover all aspects of serial I/O, but it will provide a background for understanding and selecting the interfaces covered. Only Layers 1 and 2 of the Open Systems Interconnection (OSI) model are emphasized. Higher level protocols (Layers 3–7) and some protocols are mentioned, but not covered in detail.

SERIAL VERSUS PARALLEL TRANSFER

Binary data may be transmitted from one circuit to another or from one piece of equipment to another. All bits of the words to be transmitted are sent at the same time. This requires one complete data path like a wire or PC board trace per bit. See Figure 2.1a. This is a preferred method as it is very fast. Figure 2.1b shows a more common and simpler way to illustrate parallel data buses. The one line represents multiple (*n*) lines.

However, there are disadvantages to the parallel method. First, it is more expensive as it is more hardware intensive with one connection per bit. Long connections mean multiple wire cables and related connectors. Second, as the desired transfer speed increases, bit rates on the individual lines increase converting even short data paths into formal transmission lines whose impedances must be matched to prevent reflections and distortion.

Handbook of Serial Communications Interfaces.
Doi: http://dx.doi.org/10.1016/B978-0-12-800629-0.00002-4

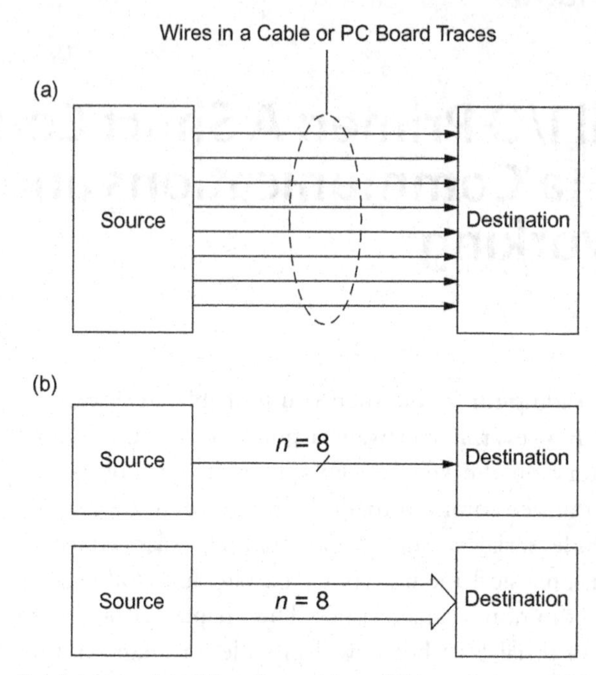

Figure 2.1 Parallel data transfers. Wires in a cable or PC board traces. (a) 8-bit parallel interface. (b) Alternate ways to illustrate parallel transfers.

Third, long lines introduce rounding distortion due to frequency limitations. In addition, capacitive and inductive coupling from closely spaced parallel lines introduces crosstalk from one path to adjacent paths. The signals transferred this way appear as noise, thereby degrading performance. Different length cable or PCB trace sizes can introduce timing skews.

With regard to integrated circuit design, parallel data transfers require more ICs pins complicating the design layout and increasing the cost of the IC packaging. With very large-scale integration, it is possible to put more and more circuitry on a chip. In addition, larger word sizes are becoming common place. For these reasons, parallel data transfers are more difficult and costly to implement.

While low-speed parallel data transfers can be made adequately, the distortion and crosstalk become prohibitive at frequencies in the tens and hundreds of megabit per second (Mb/s) ranges. Parallel path links must be limited to several feet and usually less. In the gigabit per second (Gb/s) ranges, parallel paths must be limited to only a few inches for acceptable performance.

Today, most parallel buses are very fast and connections are typically limited to short traces between ICs on a PC board or short ribbon cables with special connectors. Otherwise, data transfers are made serially.

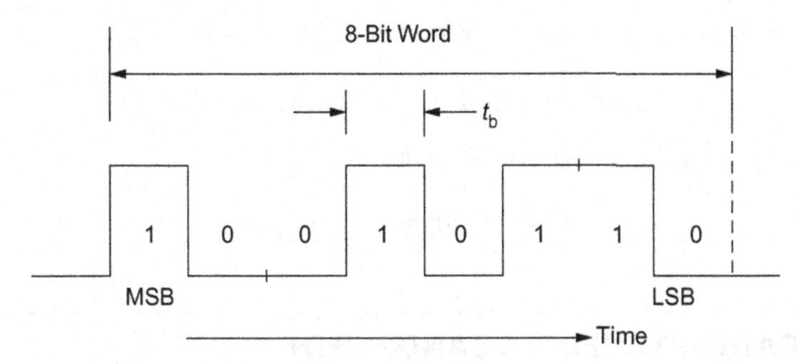

Figure 2.2 Serial data transfer.

In a serial data path, the bits of the words to be transferred are sent one at a time sequentially. The bits may be transmitted at lower sideband (LSB) first or at middle sideband (MSB) first depending on the application and interface standard. See Figure 2.2. Only a single data path (plus ground) is required. The data path is simple and less expensive.

The primary downside to a serial connection is that it takes longer to transmit a word. For example, a one-byte parallel transfer may take 50 nanoseconds (ns) over a short bus but it may take hundreds of nanoseconds serially. That may or may not be a disadvantage depending upon the usage. Serial data can easily be transmitted at Gb/s speeds over paths of several feet with some interfaces. The basic fact that there are so many different serial interfaces implies that data speed is not an issue. There is a serial interface for almost any application.

An alternative to complete serialization with one lane is to reduce the parallel lines to two or more serial lines. For example, a 64-bit bus could be reduced to 8 lines or 4 lines each carrying parts of the data in a serialized form. For example, a 32-bit word can be transferred with concurrent 8-bit serial words on four lines. Some very high-speed interfaces use this technique.

SERIAL DATA RATE

The serial data rate (R) is stated in bits per second (b/s or bps). It is the reciprocal of the bit time. The bit time is simply the time duration of a binary 1 or 0. See Figure 2.2.

$$R = 1/t_b$$

Alternately the bit time can be calculated from the rate.

$$t_b = 1/R$$

For example, a bit time of 5 ns translates to a data rate of:

$$R = 1/5 \times 10^{-9} = 200\,\text{Mb/s}$$

The bit time for a data rate of 9600 b/s is:

$$t_b = 1/9600 = 104.167\,\mu\text{s}$$

DATA RATE VERSUS BANDWIDTH

The maximum speed that a serial link can achieve is directly related to the bandwidth of the circuit or transmission medium. In most cases, the data path looks like a low-pass filter (LPF). The upper cut-off frequency determines the maximum speed of the data.

A simple approach to determining the bandwidth is to use the relationship between rise time and −3 dB bandwidth expressed as:

$$t_r = 0.35/\text{BW}$$

Here t_r is the shortest rise time and BW is the half-power bandwidth. You can also estimate the bandwidth from an observed rise time.

$$\text{BW} = 0.35/t_r$$

This provides only an estimate of the speed as using rise (or fall) times does not give a direct answer.

Another approach is to use Hartley's law that says that the channel capacity (C) or data rate in b/s is twice the channel bandwidth (B).

$$C = 2B$$

This you can see by assuming a data stream of alternating 0s and 1s as shown in Figure 2.3. The data rate is $R = C = 1/2t_b$.

This waveform is a true square wave so if all odd harmonics are filtered out by the data path, only the basic fundamental sine wave making up the square wave remains. The frequency of the sine wave is $1/2t_b$. B is the −3 dB down point of the path.

You can get an estimate of the data rate from the bandwidth. For instance, if the bandwidth is 6 MHz, the maximum data rate is:

$$C = 2B = 2(6) = 12\,\text{Mb/s}$$

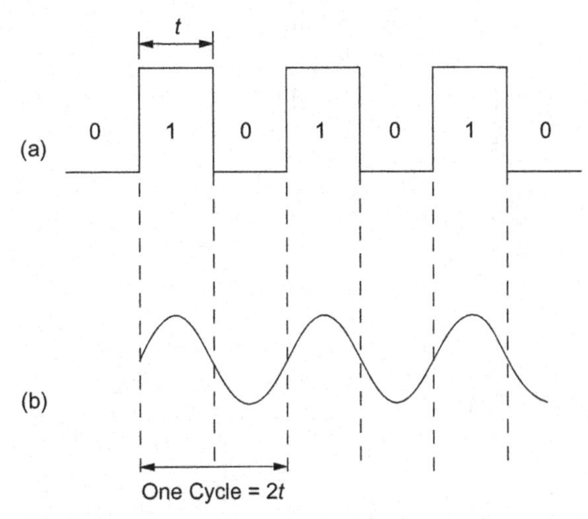

Figure 2.3 Serial data (a) and with all harmonics filtered out by the medium (b).

Keep in mind that this is just an estimate as the real data rate is also a function of the noise level in the channel. This relationship is expressed in the Shannon-Hartley theorem:

$$C = B * \log_2(1 + S/N)$$

C is the data rate in b/s, B is the channel bandwidth in Hz, S is the signal power, and N is the noise power in watts.

As a general rule of thumb, the bandwidth of a circuit or transmission medium should be five to ten times the data to allow the significant harmonics to pass thereby retaining the basic waveshape of the signal.

This relationship is also affected by the number of bits transmitted per symbol where a symbol is one of several voltage, phase or frequency variations used in a modulation scheme. Using multiple voltage levels allows more bits to be transmitted in the same bandwidth. For example, a pulse amplitude modulation (PAM) scheme called PAM4 uses four voltage levels as shown in Figure 2.4 where each level represents two bits. Each voltage level is called a symbol. The time for one symbol is the symbol rate. In this case, the bit rate is two times the symbol rate.

This discussion is for reference only and beyond the scope of this book. Most serial interfaces use standard binary two-level signals. Interfaces with multilevel symbol transmission will be noted in the coverage to follow.

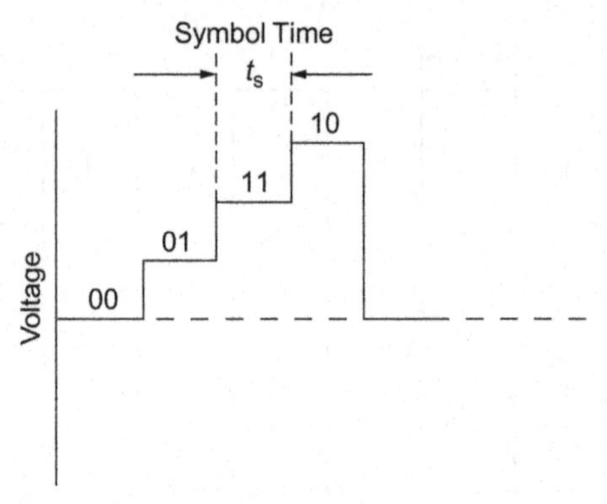

Figure 2.4 PAM4 encoding.

LINE CODING

Line coding refers to the ways that the serial data signal is shaped to represent the binary 1s and 0s. Figure 2.5 shows the most widely used line coding schemes.

Figure 2.5a shows the standard non-return to zero (NRZ) format that is the most common. The voltage level remains at a constant level for the entire bit time or does not return to zero. Most NRZ signals are unipolar, meaning both signals are of the same polarity, usually positive. TTL and CMOS levels are the most common. Some serial interfaces use a bipolar NRZ format where both positive and negative voltage levels are used. See Figure 2.5b.

A basic disadvantage of the unipolar NRZ is that the DC pulses charge the line capacitance producing an average DC component. This is unacceptable in most applications. The DC component may be removed with capacitive coupling or with a transformer. Bipolar NRZ does not have this problem as the opposite polarities average out over time.

A variation of NRZ is NRZI or NRZ Inverted. It uses a level transition to signal a binary 1 and no level transition to signal a binary 0. See Figure. 2.5c. NRZI is useful for reduced bandwidth signaling or when long strings of 1s are to be transmitted.

Another popular format is RZ or return to zero as shown in Figure 2.5d. A positive pulse represents a binary 1 that returns to zero during one bit time interval. A typical pulse width is 50% of the bit time. No pulse is transmitted for a binary 0.

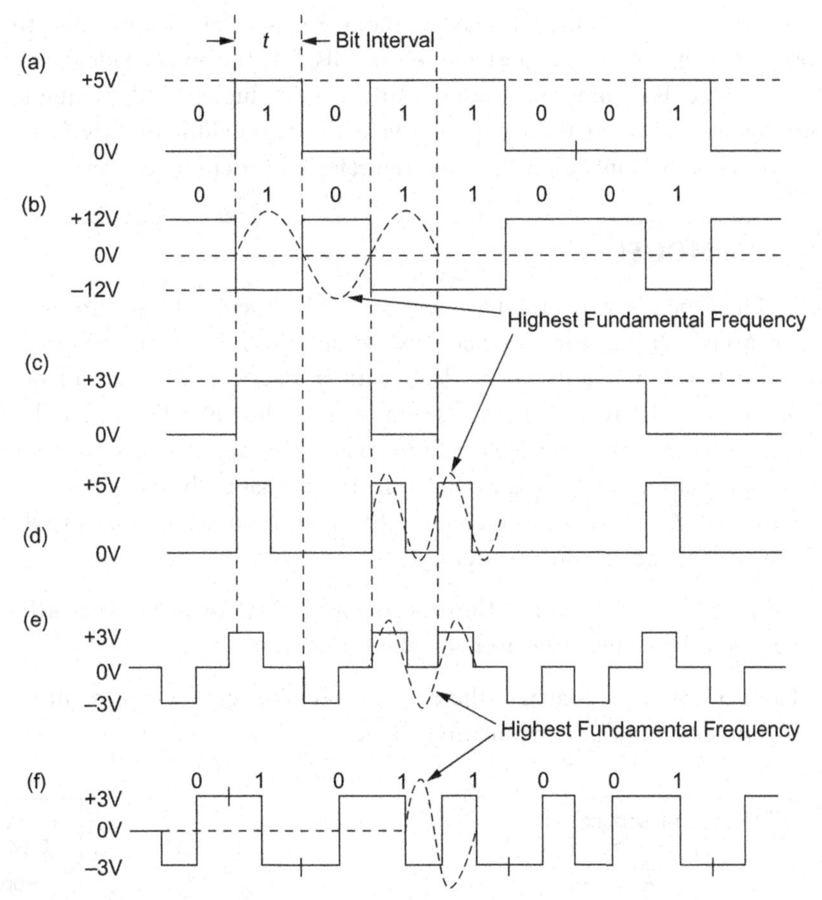

Figure 2.5 Line coding methods: (a) Unipolar NRZ, (b) Bipolar NRZ, (c) NRZI, (d) RZ, (e) Bipolar RZ, (f) Manchester.

Bipolar RZ is another option as illustrated in Figure 2.5e. A positive pulse represents a binary 1 and a negative pulse represents a binary 0. No DC build up occurs.

Manchester or biphase encoding is one more line coding method. See Figure 2.5f. It is typically bipolar, but unipolar can also be used. To represent a binary 1, a positive pulse is transmitted for one-half the bit time and a negative pulse for the remainder of the bit time. A binary 0 is transmitted as a negative pulse for half the bit time and a positive pulse for the other half of the bit time. Note that this provides a voltage transition at the center of each bit making clock recovery very easy.

There are other line coding methods for special systems but the ones shown here are the most common. NRZ is the most widely used. Methods like RZ and Manchester actually have higher (×2) frequency components, meaning they require twice the bandwidth of NRZ. The bipolar RZ and Manchester formats are preferred for clock recovery.

OSI MODEL

The Open Systems Interconnection (OSI) model is a standard of the International Organization of Standardization (ISO). It was developed as way to define how data is to be dealt with in data communications networks and systems. It is defined by seven layers as shown in Figure 2.6. The two lower layers are typically hardware while the upper layers are more commonly software. Here is a brief summary of what each layer does.

Layer 7 – Application: Implements the application and interfaces with the equipment operating system.

Layer 6 – Presentation: Ensures compatibility of data formatting between the application layer and the lower layers.

Layer 5 – Session: Manages the connection between devices by initiating, maintaining, and terminating a link.

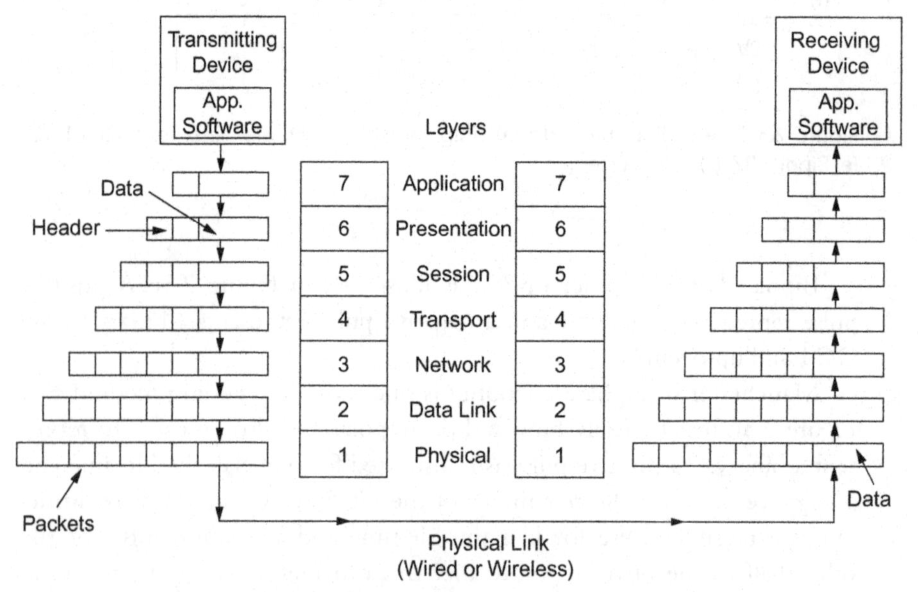

Figure 2.6 The OSI model of data communications.

Layer 4 – Transport: Ensures the quality, integrity, and authentication of the data. Flow control of the data.

Layer 3 – Network: Handles packet addressing, routing, and switching.

Layer 2 – Data Link: Defines the frame protocol, packages and disassembles the frame.

Layer 1 – Physical: Defines the transmission medium, logic levels, coding, timing, etc.

Using Figure 2.6 as a reference, follow the data from the application through the lower layers. Each layer encapsulates the data and adds a header then sends it to the next lower layer. The final packet with all the headers is then sent over the Physical link. At the receiving end of the link, the equipment then de-encapsulates the data at each layer. Finally, the application uses the data.

While most systems conform to the OSI model many do not. An example of a system that does not fit the OSI model is the Internet protocols TCP/IP. Furthermore, those systems that do follow the model typically do not use all seven layers. Layers are used as needed to deal with the protocols and application.

TOPOLOGY

Topology defines how the nodes on the network are physically connected together. The simplest connection is a peer-to-peer arrangement with just two nodes that talk to one another over the connecting medium such as a PC to printer or controller to a valve. Mostly the connections involve multiple nodes that can communicate with a central source and/or to one another.

The three basic topologies are the star, ring, and bus. Figure 2.7a shows the star where all nodes communicate with a central PC or other master embedded controller. Each node here is a transceiver. If the individual nodes want to talk to one another, they do so through the central controller. The star configuration is fairly common in networking.

The ring is a continuous closed loop of all nodes on the network. See Figure 2.7b. One of nodes acts as the main controller. Data is transmitted around the ring in one direction and all nodes see it. The ring is no longer widely used in networking.

The bus is a single cable medium to which all nodes are connected. This is shown in Figure 2.7c. The term "multidrop" is used to refer to the connection of each node to the bus. Data is bidirectional on the bus and

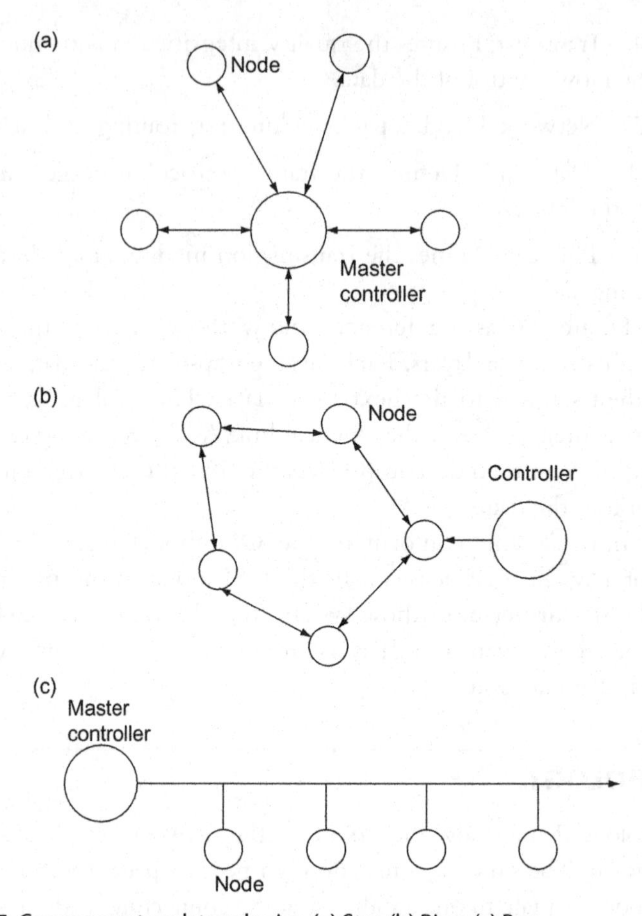

Figure 2.7 Common network topologies. (a) Star, (b) Ring, (c) Bus.

all nodes see the same data. Typically, one of the nodes serves as the master controller. The bus is the most common connection topology for most serial I/O interfaces.

BALANCED VERSUS UNBALANCED CONFIGURATIONS

Unbalanced refers to a binary voltage referenced to ground as shown in Figure 2.8a. This arrangement is simple and widely used. Only one conductor is needed along with a ground connection. However, this arrangement is susceptible to ground loop noise or external noise pick up unless the line is shielded.

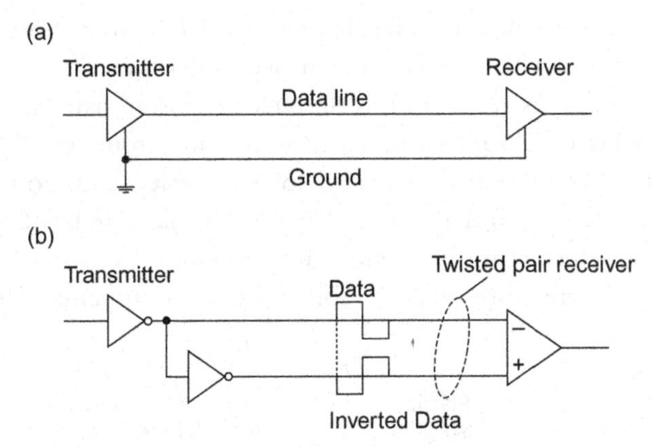

Figure 2.8 Unbalanced (a) and balanced (b) transmission.

A balanced or differential connection is shown in Figure 2.8b. The signal is transmitted on two conductors neither of which is grounded. The signal voltages on each line are of the opposite polarity with respect to ground. The signal at the receiving end is the difference between the voltages on the two lines.

The advantage of the balanced configuration is that noise common to both lines is canceled out. If equal amounts of noise voltage are induced into each line, the differential connection at the receiver subtracts it out. The primary disadvantage is that two lines instead of one are needed.

TRANSMISSION MEDIUM

The transmission medium for serial data is inherently a transmission line where the line length is 0.1 wavelength (λ) or longer at the signal frequency (f) or data rate. One wavelength is

$$\lambda = 300/f_{MHz}$$

The most common transmission lines for serial data are conductors on a printed circuit board (PCB), coax cable, or twisted pair cable. A transmission line on a PCB is one line referenced to a common ground or two parallel lines balanced with respect to ground. Such lines are short typically only several inches.

Coax cable is still used to carry serial data. Coax is an unbalanced line with a single center conductor with a solid or braided outer shielding conductor as shown in Figure 2.9a. The signal attenuation is usually high but coax can still carry signals of many Gb/s over short distances. The most common characteristic impedance is 50 ohms.

A more widely used medium is unshielded twisted pair (UTP) line as illustrated in Figure 2.9b. Two insulated solid wires from size #28 to #22 AWG are loosely twisted together to form a cable. Common examples are telephone wiring and Ethernet LAN cable like CAT5. The twisted configuration offers some protection from noise and cross talk. Shielded versions of TP are also available for noisy environments. The typical

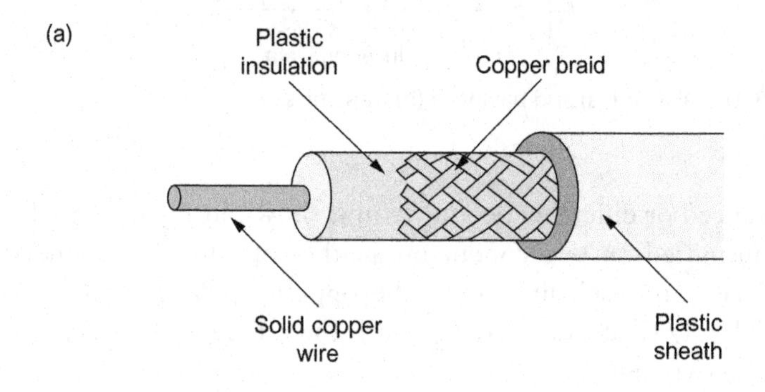

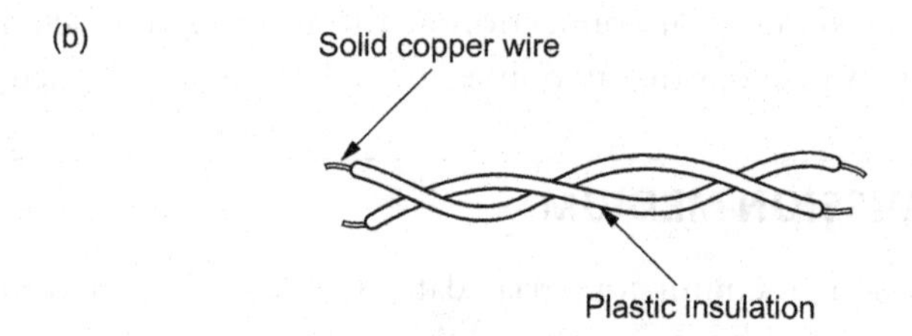

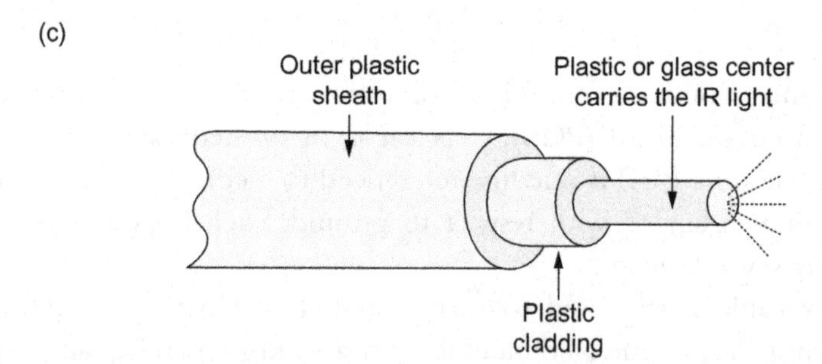

Figure 2.9 Common transmission media: (a) coax, (b) twisted pair, (c) fiber optic cable.

characteristic impedance is in the 100- to 120-ohm range. UTP is used in both unbalanced and differential configurations.

All transmission lines act as LPFs because of their distributed series resistance and inductance and shunt capacitance. As a result, any line distorts the signal. The high-frequency harmonics within the pulse signals are attenuated rounding the signal. This filtering effect depends on the inductance and capacitance per foot of cable and the length of the cable.

Another common transmission line is fiber optic cable. It consists of a glass or plastic center surrounded by a plastic cladding and outer jacket as shown in Figure 2.9c. Data is transmitted by converting the serial binary data to on-off light pulses generated by a laser or LED that are propagated down the center optical path. The light is usually invisible infrared (IR) which is used as the fiber offers less attenuation to IR than visible or UV light. The light pulses are converted back to electrical pulses at the receiving end by a light detector like an avalanche or PIN diode. Fiber optic cables are used for the fastest forms of serial data transmission. They can carry data at rates exceeding 100 Gb/s. Furthermore, they are not susceptible to electrical noise or crosstalk.

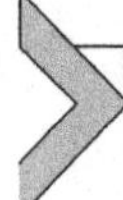

ASYNCHRONOUS VERSUS SYNCHRONOUS TRANSMISSION

Asynchronous serial data is usually sent one binary word at a time. Each is accompanied by start and stop bits that define the beginning and end of a word. Figure 2.10 shows one byte with a start bit that transitions from 1 to 0 for one bit time. The byte bits follow. A 0 to 1 transition or a

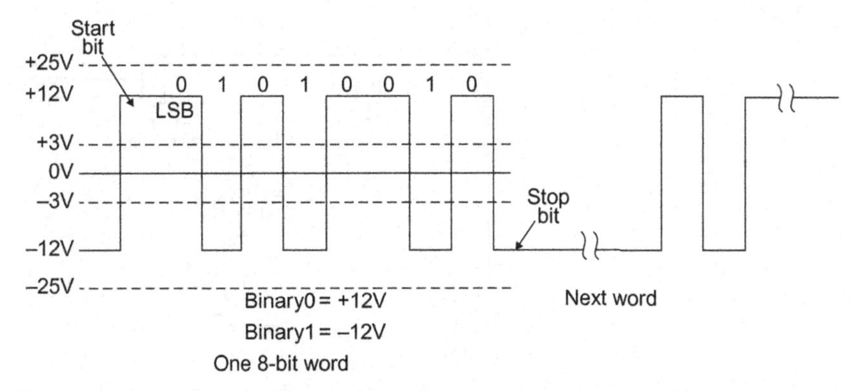

Figure 2.10 Asynchronous data transmission.

low signal is the stop bit the end of the word. Some standards use two stop bits. Note that a 1 is low (−12V) and a 0 is high (+12V).

Asynchronous transmission is very reliable but is inefficient because of the start and stop bits. These bits also slow transmission. However, for some applications, the lack of speed is not a disadvantage where reliability is essential.

Many serial interfaces and protocols use this asynchronous method. It is so widely used that the basic process has been incorporated into a standard integrated circuit known as a universal asynchronous receiver transmitter or UART. Figure 2.11 shows the basic arrangement. Data to be transmitted is usually derived from a computer or embedded controller and sent to the UART over a parallel bus and stored temporarily in a buffer memory. It is then sent to another buffer register over the UARTs internal bus. The byte to be sent is then loaded into a shift register where

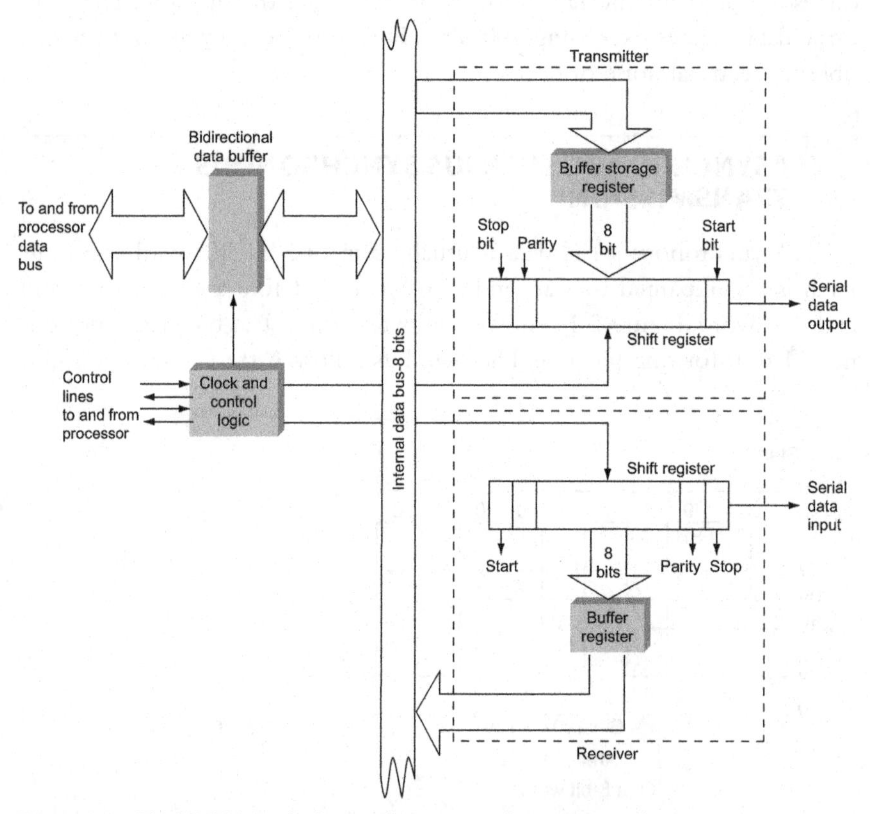

Figure 2.11 The UART is at the heart of most asynchronous interfaces.

start, stop, and parity bits are added. Parity is usually optional and odd or even parity may be selected. A clock then transmits to word serially in NRZ format.

Received data comes in serially to the lower shift register. The start, stop, and parity bits are monitored and actions are taken accordingly. The received byte is then transferred in parallel to the buffer and then to an external microcontroller or a PC. Control logic provides input and output control signals used in some protocols. While UARTs are available as an IC today, they are more likely to be integrated into an embedded controller or other IC.

The overhead of the stop and start bits can be eliminated or at least minimized by using synchronous transmission. Each data word is transmitted directly one after the other in a block where the beginning or end is defined by selected bits or words. Digital counters keep track of word boundaries by counting bits and words. Synchronous transmission is faster than asynchronous because there are fewer overhead bits. However, synchronous transmission does have its overhead. Data is transmitted in blocks or frames of words and special fields define the beginning and end of each frame. These codes take the form of multibit words that form the packet overhead.

The term "asynchronous" also refers to the random start of any transmission. In most data communications systems, data is transmitted in packets, usually synchronous. However, it is not known when a packet will begin or end as it is not tied to any timing convention. Data is just sent when needed.

CLOCK AND DATA RECOVERY

The circuitry at both the transmitting unit and the receiving units requires a clock signal. There are three basic ways that the clock is provided. In the first method, each transmitting and receiving units have their own clock sources. The clocks are equal but not synchronized. The format of the data and the timing is such that small differences in clock frequency can be tolerated. Asynchronous data transmission works like this.

A second method is for the transmitter to provide a clock signal to the receiver over a separate wire. Some serial interfaces use this method. The disadvantage is that a path separate from the data path is needed. The advantage is perfect synchronization.

A third method is for the receiver to extract the clock from the transmitted data. Most data signals have a sufficient number of transitions that

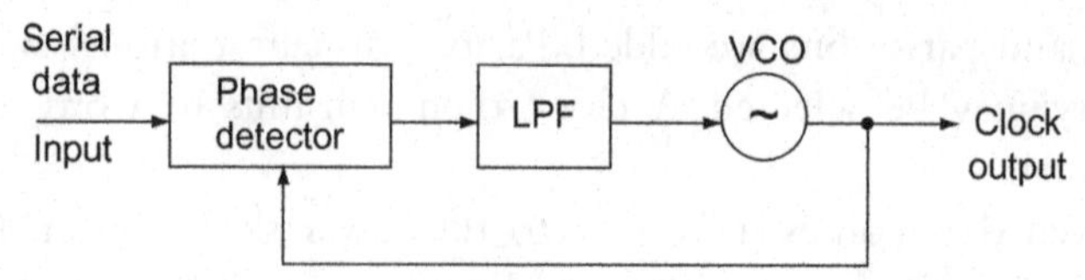

Figure 2.12 A phase-locked loop (PLL) clock recovery circuit.

receiving units can use to reconstitute the clock. This is called clock and data recovery (CDR).

In order for the receiving circuitry to stay in sequence with the transmitting circuits, the two clocks driving the circuits must be at exactly the same frequency. This is difficult to achieve with separate clocks even if crystal oscillator precision and tolerance is used. While some low-speed asynchronous circuits can function correctly with some clock discrepancy, high-speed data circuits cannot. The solution to this problem is simply to derive the receiving clock from the transmitted data transitions.

A number of techniques have been developed over the years to recover the clock, today the single most widely used technique is to use a phase-locked loop (PLL). The basic circuitry is shown in Figure 2.12. The serial data is signal conditioned and applied to one input of the PLL phase detector. The clock output from the voltage-controlled oscillator (VCO) is applied to the other phase detector input where the two are compared. If any frequency or phase difference occurs, the phase detector produces an error signal that is converted into a DC voltage by the LPF. This DC drives the VCO to change its frequency to match the input. The result is a clock signal whose frequency matches the incoming data rate.

This technique works well but to function there must be a sufficient number of binary 0/1 and 1/0 transitions in the input to keep the PLL in the locked condition. In addition, even if there are alternating 1s and 0s, the resulting input frequency is one-half the bit rate since there is one bit per half cycle of signal. What is needed is a clock frequency that is one clock pulse per bit. To compensate for this, a flip flop or divide by 2 circuit is placed between the VCO output and the phase detector input. This causes the PLL to act as a ×2 multiplier giving the correct clock frequency or one clock pulse per input bit. The clock signal is then further conditioned as needed and then used to drive the other circuits as needed.

A variety of other techniques are used with the PLL including type of phase detector, edge detection, and line coding method (RZ vs. NRZ vs. Manchester, etc.), as well as data encoding, to ensure accurate clock recovery.

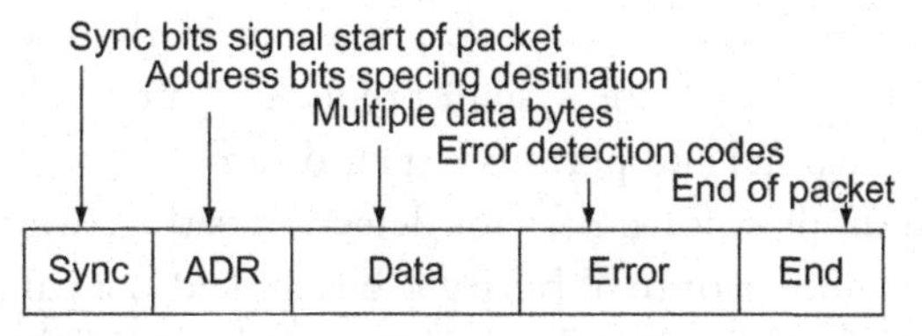

Figure 2.13 Synchronous transmission does not use start and stop bits but transmits data within a packet or frame with identifier fields making up the protocol.

PROTOCOLS

Protocols are rules and procedures that define how nodes on a network communicate. Protocols may be signals or messages passed between communicating entities to ensure a reliable transmission. Each interface type has its own specific guidelines. Along with these procedures is the definition of how the data to be transmitted is packaged. Most data is sent in the form of blocks or frames that define how many bytes are to be transmitted and in what sequence and format.

All protocols define a frame format that consists of fields of bits that designate the start and end of the frame, destination and/or source node addresses, number of bytes sent, error correction type, and other information relevant to the transmission.

A typical frame format is shown in Figure 2.13. It begins with a byte or longer word that signals the start of the frame that the receiver recognizes as such. It initiates clock synchronization in the receiver. Next is a field of address bits that identifies the receiving node. Next is the data usually a large number of bytes. Following the data is usually a field containing the error correction code. Finally, the remaining bits signal the end of the frame. All other frame formats follow this generic sequence with special fields and bits added to implement the unique protocol. Frames may be transmitted asynchronously one byte at a time (with start and stop bits) in sequence. Alternately, the frames can be transmitted synchronously.

ERROR DETECTION AND CORRECTION

Many of the fastest serial interfaces incorporate protocols that include some form of error detection or forward error correction (FEC). These techniques improve transmission reliability and lower the bit error rate (BER). The FEC methods are so good that in the presence of noise they actually act as though the system produced gain. Coding gain offsets the noise in the channel and permits higher data rates.

Parity. The simplest form of error detection is the parity bit. Circuitry determines the number of total binary 1 bits and adds a parity bit indicating where an odd or even number of bits is to be transmitted. The parity bit is added to the word to be transmitted. At the receiver, again circuitry determines the number of odd or even bits and compares the result to the received parity bit. If there is a difference, an error occurred. However, this method does not determine where or what the bit error is. An error usually initiates a retransmission of the word or frame.

Block Check Code. A block of data to be transmitted is subjected to a complex algorithm that computes a multibit word that is transmitted along with the data. One such method creates a block check code (BCC) or block check sequence (BCS). A sequence of words to be transmitted are all added or XORed to produce a single BCC character that is transmitted at the end of the data. The process is repeated at the receiver and the check words are compared. Versions of this method permit identification of specific bit error locations.

CRC. A commonly used error detection method is the cyclic redundancy check (CRC). It is a mathematical method that divides a block of data by a constant. The quotient is discarded and any remainder is transmitted as the CRC. The CRC is usually implemented in hardware with shift registers and XOR gates. It can have different length words usually 8, 16, or more bits. The CRC is generally capable of detecting up to 99.9% of all bit errors. CRC does not correct the errors.

FEC. FEC methods use more elaborate methods. One popular example is the Reed Solomon FEC. In one configuration, a block of 255 bytes is processed producing a 32-bit parity word that is appended to the data. At the receiver, the received data and the FEC parity word are processed to determine just what bit failed and then corrects it.

There are many types of FEC and they are beneficial as they greatly improve data integrity and reliability. Examples are convolutional and turbo codes that are beyond the scope of this book. However, there is one key point to consider. The FECs do add many extra bits to the data slowing data transmission. Usually, higher clock speeds are used during transmission to ensure that a target real data rate is achieved. For example, 10 Gb/s Ethernet has a data transmission line rate of 10.3125 Gb/s that will create a true even 10 Gb/s data rate.

ACCESS METHODS

When more than two devices share a common transmission medium, some means must be provided to determine who uses the medium and when. These are called access methods. For most of the interfaces covered in this book, one of three methods are used. These are master-slave, carrier sense multiple access (CSMA), and time division multiple access (TDMA).

Master-Slave. In a master-slave access system, a single master controls two or more slaves. The master effectively tells the slave when to transmit or receive so there are no conflicts on the common bus medium. Interface protocols use commands to instruct the slaves. All operations are determined by the master. A common technique is polling of the slaves in sequence sending or receiving data as needed.

CSMA. Carrier sense multiple access (CSMA) is a system that causes all nodes on a bus to listen before transmitting. If data is being transmitted, it means that a "carrier" is present so do not transmit. Otherwise it will cause a collision or interference where both signals will mix and cause massive errors. Most CSMA systems also incorporate a collision detection (CD) technique that identifies any collisions. A collision causes all parties to stop transmitting and to wait for a random amount of time before trying again. CSMA/CD automatically handles any contention on the bus. This extra contention for the bus reduces the overall throughput and increases latency.

TDMA. Time division multiple access (TDMA) uses a time allocation method to allow multiple users to share a common medium. The system defines a unit of time for one cycle. This time is divided up into time slots, one for each of the nodes on the shared medium. When a node needs to transmit or receive data, it waits for its time slot and then initiates the communications. Time slots are often one word or byte long but could be longer. The speed of transmission is greatly affected by this method, but in most systems where it is used, this added time is not a disadvantage because of the overall data rate.

DUPLEXING

Duplexing refers to two-way communications. Half duplexing means alternating transmit and receive operations on a single channel or line. Both parties cannot transmit at the same time.

Full duplex means simultaneous transmit and receive operations. In wired systems, it means that two channels or lines are needed. When a single line is used, typically full duplex operation is achieved with TDMA methods.

Simplex means one-way operations only. It is usually a broadcast mode to multiple nodes.

MISCELLANEOUS SERIAL TECHNIQUES

One problem that often occurs in transmission is a long sequence of binary 0s or 1s. When this occurs, the CDR circuits in the receiver lose lock and the VCO operating frequency can change producing clocking errors. One way to overcome this problem is to use what is called 8b/10b conversion or coding. This is a technique that translates each 8-bit sequence of bits into a 10-bit sequence. The coding is such that there is always a minimum number of binary 1 bits per byte, enough to keep the CDR PLL in lock.

The 8b/10b encoding also helps provide a zero DC balance since it ensures a 50/50 percent mix of binary 1s and 0s. The coding also enables a method to keep the data words aligned. The dividing line between sequentially transmitted bytes or longer words is lost during transmission but the coding allows circuits to determine the word delineation.

The 8b/10b coding is very common but it also represents a 25% overhead that reduces the data rate. This amount of overhead is fine for lower speeds but for high-speed systems a similar technique called 66b/64b is used. Here the overhead is only 3.125% and it too provides the desired DC balance and word delineation.

Another technique is bit scrambling. This technique uses linear feedback shift registers to produce a pseudo-random bit sequence that is useful in some applications. It can provide a better distribution of 1s and 0s to overcome long sequences of 1s and 0s that cause the receiver CDR to lose lock. It helps the DC balance problem and is useful in wireless applications where the data is used to provide automatic gain control (AGC) or equalization. Bit scrambling can also help keep the power spectral density of modulated signals within a specific range as required by standards. The signal received is descrambled by another linear feedback shift register.

Bit interleaving is one more technique sometimes employed in serial I/O applications. It is used in conjunction with FEC to reduce burst errors. In some applications, wireless especially, a random burst of noise can create a bit error or sequence of bit errors. By storing the words to be transmitted and the related FEC then sending them one bit at a time from each word, it is possible to identify bit errors at the receiver and correct them.

These techniques are often used in combination with FEC to produce a desired outcome depending on the application.

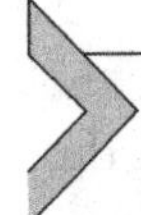

SERIAL TO PARALLEL AND PARALLEL TO SERIAL CONVERSION

To transmit data serially it is usually necessary to convert it from a parallel form from a storage register or bus into a serial bit steam. This is called parallel to serial conversion or serializing. At the receiver, the serial data is then usually converted back to parallel from processing or storage. This is accomplished by serial to parallel conversion or deserializing. Circuits for doing this conversion are often called serdes or SERDES short for serializer/deserializer.

Serdes operations are typically carried out by shift registers. Figure 2.14 shows how parallel data is loaded into a shift register and the bits are shifted

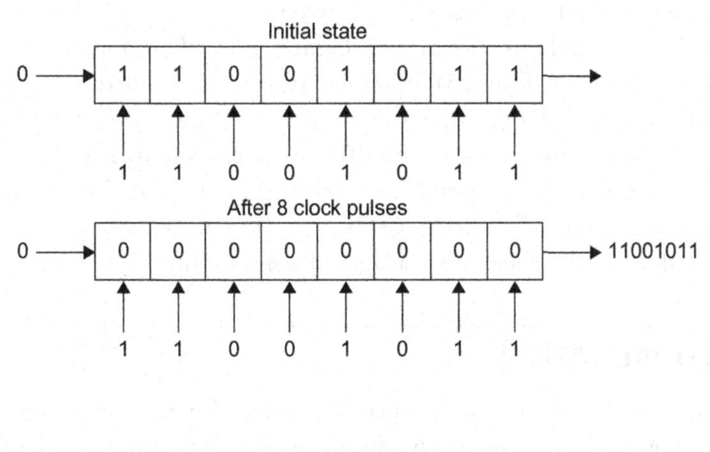

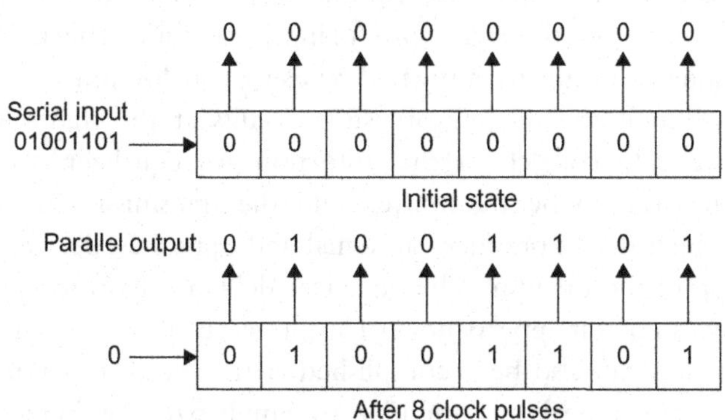

Figure 2.14 Parallel-to-series conversion (a) and series-to-parallel conversion (b).

out one bit at a time by the clock pulses. The reverse process is carried out by shifting the serial data into a shift register one bit at a time, transferring the shift register contents into a storage register on to a parallel bus.

The key to successful serdes operations is synchronized clocks at the transmitter and receiver. The CDR circuitry described earlier ensures this proper synchronization. All of this circuitry is typically available in a single integrated circuit that usually includes the CDR function.

GEARBOX OPERATIONS

"Gearbox" is a term applied to circuits that translate serial data streams at one speed to a serial stream at a higher or lower data rate. The basic circuitry is made up of multiple shift registers, multiplexers/demultiplexers and related logic along with multiple clocks that provide the timing. One application is cable or path reduction. For instance, eight serial sources can be aggregated into a single higher-speed serial signal and vice versa with 8:1 and 1:8 gearboxes, respectively.

Another example is shown in Figure 2.15. This circuit converts 10 Gb/s streams into four 25 Gb/s data streams. It is called a 10:4 gearbox. Another circuit performs the reverse function of converting the four 25 Gb/s streams back into ten 10 Gb/s streams. This is called a 4:10 gearbox. Such circuitry is useful in translating between different forms of the Optical Transport Network (OTN) or 100 Gigabit Ethernet systems. These serdes-based devices are available in a single chip.

EQUALIZATION

All serial data media distort the signal being transmitted. The cables act as LPFs that effectively eliminate or at least attenuate the high-frequency content of the signal producing a rounding effect. This effect can be eliminated or greatly mitigated by using equalization.

Equalization is the process of using circuits at the transmitter and/or the receiver to correct for the distortion. An equalizer is connected between the data to be transmitted and the transmission medium. Its response is such that it produces an equal and opposite effect to the low-pass filtering of the medium. The equalizer deliberately distorts the transmitted signal so that it appears normal at the receiver.

Equalization can also be accomplished at the receiver. A typical technique is to use a peaking amplifier to emphasize the high-frequency

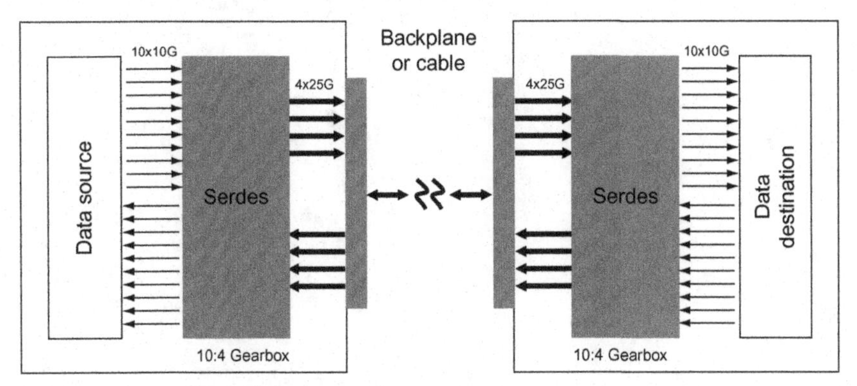

Figure 2.15 10:4 and 4:10 Serdes gearboxes.

content of the signal to restore its original waveshape. In some cases, equalization is used at both the transmitter and receiver.

Several different methods are used including feed forward equalization (FFE) and decision feedback equalization (DFE), both are beyond the scope of this book.

BIT BANGING

A term you will sometimes hear related to serial I/O is "bit banging." This refers to the process of implementing serial I/O procedures in software in an embedded microcontroller. It is most often associated with the asynchronous transmission method used with a UART interface. However, the term can apply to any other serial protocol implemented in software rather than in a specific interface IC. If a microcontroller does not have the desired integrated interface, it can be programmed using the processor I/O pins. Detailed knowledge of the protocol is needed to get the desired sequences, formats, and timing correct.

FIELDBUSES

One of the terms you will see with reference to serial interfaces is fieldbus. A fieldbus is just a serial interface and network for connecting field instruments such as sensors, transducers, and actuators like valves, relays, solenoids, and pneumatic/hydraulic cylinders. These are used in process control systems and factory automation. Fieldbuses connect directly to the controller or computer that operates them. Fieldbuses are typically bridged to a higher level network such as Ethernet in larger systems.

CHAPTER THREE

Selecting an Appropriate Interface

With dozens of serial interfaces available, choosing the right one for a specific design would seem to be difficult. The process of selecting an optimum interface is made easier with this book. Here are a few guidelines to help your selection process.

1. If your design calls for some specific interface like universal serial bus (USB) or serial peripheral interface (SPI), then your choice is simple. Often this interface choice is provided for you as the embedded controller or other ICs used in the design may already have on-chip interfaces ready to use. Usually an easy choice.
2. If your design must connect with some external equipment, consider what that might be and determine the existing or potential interfaces.
3. Consider the environmental considerations. Will the equipment be used in a harsh environment or noisy location? Some interfaces are better than others in such environments. You may want the noise immunity of a differential connection instead of the simpler single-ended connection. High noise levels may also rule out wireless as an option.
4. What speed is necessary? Consider the time it takes to send the data and any latency issues. Determine any need for real-time determinism. Interfaces with speed capability of a few bits per second to 100 Gb/s are available and everything in between. The wired interfaces in this book are divided into three groups by speed: 0 to 10 Mb/s, 10 Mb/s to 1 Gb/s, and 1 Gb/s to 100 Gb/s. Wireless interfaces tend to be slower in terms of bit rate and latency.
5. Consider the transmission path length. Is the link just on a PC board or many meters away? How many pins are cable wires are available? These factors will influence the type of transmission medium and speed. Wireless can eliminate the cabling problem in some applications.
6. Is data reliability critical? If so, will error detection and/or correction be needed to ensure communications integrity? Will a simple parity

Handbook of Serial Communications Interfaces.
Doi: http://dx.doi.org/10.1016/B978-0-12-800629-0.00003-6

check be sufficient or will advanced forward error correction (FEC) be needed?

7. Is clock timing an issue? Is a separate clock line possible for solid clocking or will the clock be derived from the data signal? Is jitter an issue?
8. Are integrated circuits or modules available to implement the interface? If so, it will save time and money in development. If not, then a more time-consuming and costly custom design will be necessary. This is a critical point as it will affect the design time. This is a build versus buy decision. If complete interface modules are available, design time will be less. Designing from scratch with available ICs will take time.
9. Cost is always an issue. This may be a prime consideration so evaluate it as you assess each interface choice.
10. Interoperability may be an issue in a design. If your equipment must work with other interfaces and be compatible, be sure to choose an interface that will have the desired compatibility.
11. Internetworking is another consideration. Will the interface ever have to connect to another network like Ethernet or the Internet, be sure this is some compatible or potential for data exchange.
12. Go with what is best for your application. Choose an interface that is the one most likely to make your design a success.

Once you have this information, go to the section of this book describing interfaces in the data speed group you selected. Read through the list of interfaces listed and then zero in on the best one for you.

When you have selected the interface, go to the organization supporting that interface or standard and acquire all the related publications and documents. Search for any available books elaborating on the interface. Also survey the integrated circuit manufacturers to see what chips are available. Be sure to look for related application notes, FAQs, data sheets, white papers, etc. It wouldn't hurt to do an Internet search to locate any useful application information. Finally, check with the test instrument manufacturers to see if test information is available. Several instrument companies offer equipment and software to make testing some interfaces faster and easier.

IMPORTANT NOTE

The interface descriptions in this book are summaries that boil down the details you need to make an interface decision. Only the physical (PHY) layer (Layer 1) and Media Access Control (MAC) layer (Layer 2) are considered. Most interfaces will have higher level layers that you will consider after selecting the basic interface type. Protocol highlights are included but very little detail because of their complexity.

SECTION I

Low-Speed Interfaces (0–10 Mb/s)

CHAPTER FOUR

1-Wire

APPLICATIONS

- Reading remote temperature sensors
- Data logging (temperature, humidity)
- Simple remote control (off-on), switching
- Device identification (ID)
- Memory (NV SRAM, OTP, EEPROM)
- Timing/counting, real-time clock
- Access control
- Secure authorization and ID

SOURCE

Dallas Semiconductor now part of Maxim Integrated.

NATIONAL OR INTERNATIONAL STANDARD

None

KEY FEATURES

- Low speed
- Long range
- Simplicity
- Low cost
- Parasitic power for slaves
- Available ICs or embedded in microcontrollers or selected devices.

DATA RATE

Approximately 16.3 kb/s (maximum). Depends on the length of bus and number of connected nodes.

Handbook of Serial Communications Interfaces.
Doi: http://dx.doi.org/10.1016/B978-0-12-800629-0.00004-8

CABLE MEDIUM

- Twisted pair like telephone wire or Ethernet CAT5/6/7 cable
- Flat ribbon cable
- Modular telephone cable (RJ-11 connectors)

RANGE

Up to about 300 meters. Depends on the number of slaves connected.

NETWORK CONFIGURATION

- See Figure 4.1 Multidrop bus.
- Name of network: MicroLan.
- Single-ended, unbalanced 1 wire plus ground bus.
- Open drain master controller and one or more slave nodes. See Figure 4.2.
- Half duplex, bidirectional.
- Maximum of 63 slave devices.
- Each slave contains a single rectifier diode and storage/filter capacitor. The capacitor stores the line (binary 1) voltage from the master and filters the pulses on the line into a DC supply voltage for the slave device circuits.
- Each slave device contains an 8-byte/64-bit ROM to store an address ID code. See Figure 4.3. The first byte is an IC family code. The next six bytes are the unique device serial number. The last byte is an 8-bit cyclic redundancy check (CRC) code based on the first 7 bytes. Transmission is lower sideband (LSB) first.

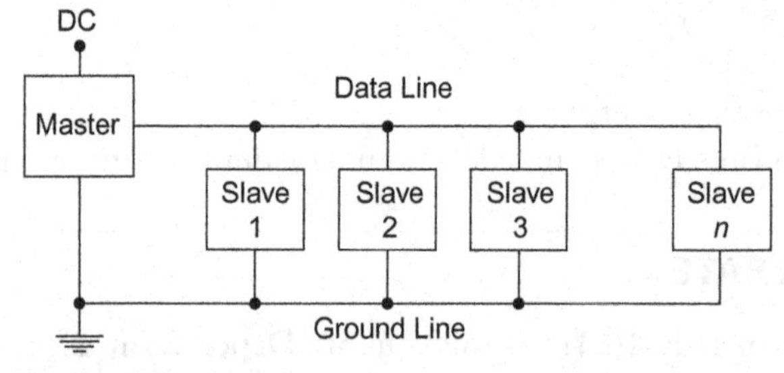

Figure 4.1 1-Wire bus (MicroLan)

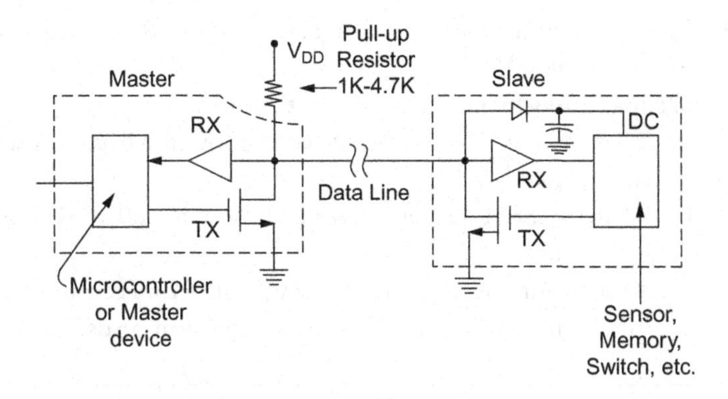

Figure 4.2 1-Wire bus with Master and Slave details showing transmitter (TX) MOSFETs and receivers (RX).

8-Bit CRC	48-Bit (6-Byte) Serial Number	8-Bit Family Code
MSB		LSB

Figure 4.3 64-bit (8-byte) slave ID code.

LOGIC LEVELS

- CMOS/TTL compatible
- Binary 0 (low): 0 to 0.8V, Binary 1 (high): 2.2 to 5 volts.
- Pull-up drain resistor needed external to the master device.

PROTOCOL

1. Master device sends binary 0 reset pulse for 480 μs to reset the system.
2. Master device waits for slave(s) to send a binary 0 presence pulse for at least 60 μs.
3. The master asks for the slave address with an 8-bit command. Slave responds.
4. An 8-bit command byte is sent indicating the function the slave is to perform. Commands include directions to slaves to select/enable, ignore, read, write, match/compare memory, switch, real time clock interrupt, etc.

5. Slave then either transmits or receives data in 8-bit segments followed by the CRC.
6. Write/Read Operations
   a. Write binary 1 to slaves: Drive bus low for 6 μs. Release bus, wait 64 μs.
   b. Write binary 0 to slaves: Drive bus low for 60 μs. Release bus, wait 10 μs.
   c. Read from slaves: Drive bus low, wait 6 μs. Release bus, wait 9 μs. Sample bus to read bit from slave, wait 55 μs.

IC SOURCES

- Atmel
- Maximum Integrated
- Texas Instruments

Actuator Sensor Interface (AS-i)

APPLICATIONS

- Factory automation
- Industrial and process control

SOURCE

Developed by Siemens and managed by the AS International Association.

NATIONAL OR INTERNATIONAL STANDARD

- EN 50295
- International Electrotechnical Commission (IEC) 62026-2

KEY FEATURES

- Common nodes are pushbuttons, keyboards, limit switches, indicators, relays, solenoids, valves, indicators and other on–off devices.
- Simplicity.
- Low cost.
- Flexibility.
- Can be used alone or as an extension or supplement to another interface or field bus.
- DC power (24V) and signal on the same cable.
- Used primarily with programmable logic controllers (PLCs) for on–off monitoring and control.
- Support for analog I/O.

Handbook of Serial Communications Interfaces.
Doi: http://dx.doi.org/10.1016/B978-0-12-800629-0.00005-X

DATA RATE

167 kb/s (6 μs bit time).

CABLE MEDIUM

Special yellow flat non-twisted, non-shielded #16 or #14 AWG stranded. Impedance 70 to 140 ohms: termination needed. Connections made by special connectors that pierce the insulation to make contact with the wires. Options include a similar flat black cable for DC power only and a round cable for special applications.

RANGE

100 meters (328 feet). Up to 300 meters with repeaters.

NETWORK CONFIGURATION

- Flexible topology: may be a bus, daisy-chain, ring, star, tree, etc.
- Balanced differential line.
- Master-slave access, one master and up to 31 or 62 slaves.
- Master polls slaves in a 5 ms (10 ms with 62 slaves) cycle.
- Each slave data packet has 4 input bits and 3 output bits. With 62 slaves, a total of 248 input devices and 186 output devices can be accommodated.
- Parity error detection with retransmission.

LOGIC LEVELS

Data in non-return to zero (NRZ) form is encoded in Manchester format and differentiated by an inductor to form alternating pulse modulation (APM). The APM signal pulses are ac coupled to the cable and superimposed on the dc. Typical levels are ±2V of the DC line voltage. See Figure 5.1.

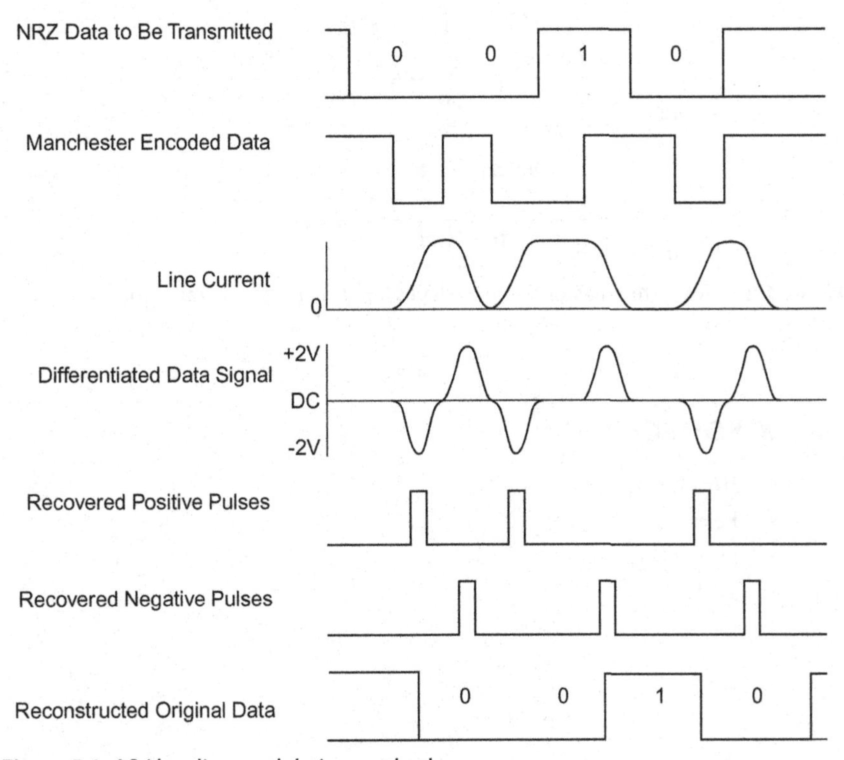

Figure 5.1 AS-i bus line modulation method.

PROTOCOL

- Master sends out 14-bit call to slaves: one start bit (SB), one control bit (SB), 5-bit address (A0–A4), 5-bit command or information word (D0–D4), one parity bit (PB), one end bit (EB). Figure 5.2a.
- All slaves are polled in sequence.
- Each slave can support 4 input bits and 3 output bits.
- Slave responds with 7-bit frame with one start bit (SB), four information bits requested by master, one parity bit, and one end bit. Figure 5.2b.

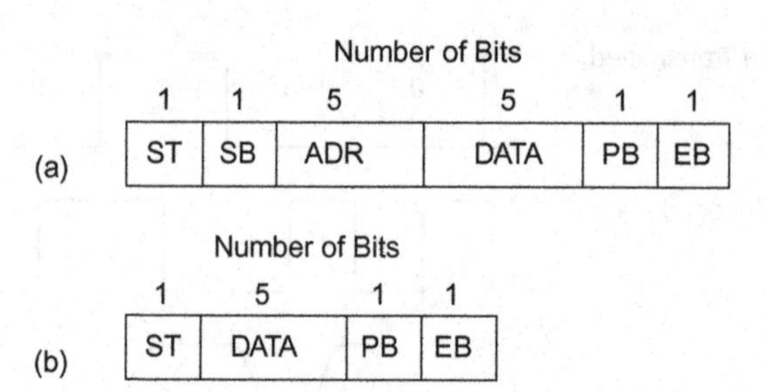

Figure 5.2 Transmitted master frame to slaves (a), slave response frame (b).

IC SOURCES

- Bosch
- Festo
- Siemens
- ZMDI

CHAPTER SIX

BITBUS

APPLICATIONS

Industrial control and factory automation.

SOURCE

Intel Corporation was the original developer of BITBUS.

NATIONAL OR INTERNATIONAL STANDARD

Institute of Electrical and Electronic Engineers (IEEE) 1118.

KEY FEATURES

- One of the oldest field buses.
- Microcontroller based, similar to the Intel 8051 family. Original IC is the i8044.
- No longer widely used in the United States, but still popular in Europe.

DATA RATE

62.5 kb/s and 375 kb/s self-clocked and from 500 kb/s to 2.4 Mb/s synchronous.

CABLE MEDIUM

Two pairs of unshielded twisted pair; one pair for data and the other for the clock.

Handbook of Serial Communications Interfaces.
Doi: http://dx.doi.org/10.1016/B978-0-12-800629-0.00006-1

RANGE

Depends on data rate and number of nodes. Up to 1.2 km at 62.5 kb/s or 300 meters at 375 kb/s. Maximum range estimated to be 13.2 km.

NETWORK CONFIGURATION

- Uses standard RS-485 physical layer.
- Multidrop bus.
- Master-slave architecture.
- Up to 250 nodes.

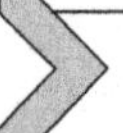

LOGIC LEVELS

Same as RS-485. Uses NRZI encoding.

PROTOCOL

- Uses the IBM developed Synchronous Data Link Control (SDLC) protocol. See Figure 6.1.
- Address and control bytes designate the node to send or receive data and how the data should be handled.
- Data block length to 248 bytes.
- 16-bit CRC.
- Flags signal start and end of a frame.

Number of Bits per Field

8	8	8	0-248	16	8
Start flag	Address	Control	Data	CRC	End Flag

Figure 6 1 The BITBUS protocol frame is similar to the IBM-developed synchronous data link control (SDLC) protocol.

IC SOURCES

Rochester Electronics

CHAPTER SEVEN

C-Bus

APPLICATIONS

- Home networking and automation (control of lighting, HVAC, security, etc.)
- Commercial building automation

SOURCE

Developed by Clipsal Integrated Systems in Australia. Now integrated with Schneider Electric.

NATIONAL OR INTERNATIONAL STANDARD

None

KEY FEATURES

- A higher layer protocol.
- Uses RS-232 for PC interfaces.
- DC power for interface units carried over the data cable (36V).

DATA RATE

Typically 9600 b/s.

CABLE MEDIUM

CAT5 unshielded twisted pair.

Handbook of Serial Communications Interfaces.
Doi: http://dx.doi.org/10.1016/B978-0-12-800629-0.00007-3

CONNECTORS

- DB9 for RS-232
- RJ-45

RANGE

1000 meters maximum. Greater length using network bridges.

NETWORK CONFIGURATION

- Peer-to-peer or full network.
- Point-to-multipoint configuration.
- Daisy-chained or bus connections.
- Maximum number of nodes 255.
- Uses RS-232 to connect interfaces to a PC.

LOGIC LEVELS

Compatible with RS-232.

PROTOCOL

- Asynchronous using RS-232 standards.
- Sequence of byte length addresses and commands.
- Example of lighting control: Input switch is operated generating an ON command. Command is sent over network to all units programmed to be affected. Lights programmed to be operated turn on, others on the network ignore the command.

IC SOURCES

- See IC sources for RS-232.
- Primary equipment source Clipsal Integrated Systems and Schneider Electric.

CHAPTER EIGHT

Controller Area Network (CAN)

APPLICATIONS

- Automotive control systems
- Industrial automation

SOURCE

Originally developed by Robert Bosch GmbH in 1983 and sanctioned by the Society of Automotive Engineers (SAE) in 1986.

NATIONAL OR INTERNATIONAL STANDARD

International Standardization Organization (ISO) ISO-11898, ISO-11519 and others.

KEY FEATURES

- Low cost
- Light weight
- Excellent noise immunity
- Priority data handling
- Real-time data handling
- Error checking for reliability
- Many IC suppliers

DATA RATE

5 kb/s (minimum), 10, 20, 50, 125, 250, 500, 800 kb/s, 1 Mbps

CABLE MEDIUM

Shielded or unshielded twisted pair or ribbon cable.

Handbook of Serial Communications Interfaces.
Doi: http://dx.doi.org/10.1016/B978-0-12-800629-0.00008-5

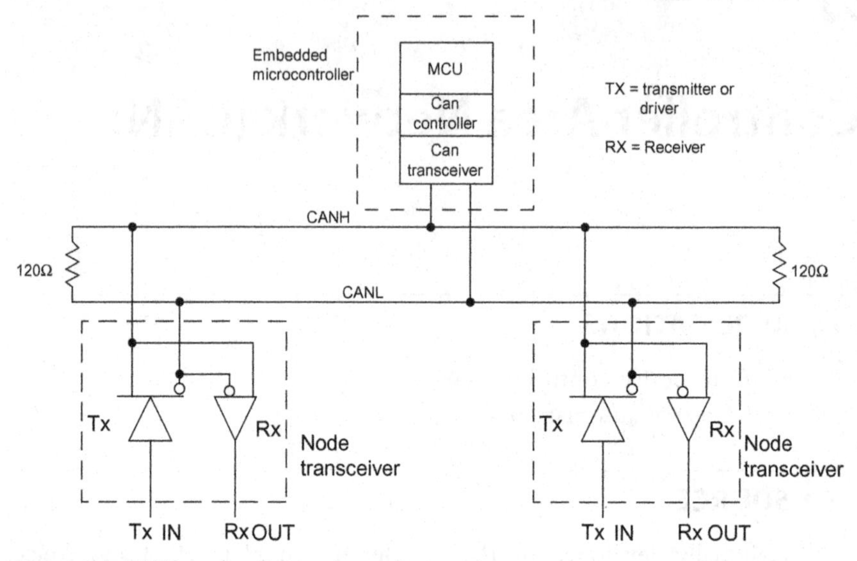

Figure 8.1 CAN bus showing nodes and transceivers.

CONNECTOR

DB9 (male)

RANGE

1000 m maximum (40 kb/s). Depends on the bus length and number of nodes, e.g., 1 Mb/s up to 40 m.

NETWORK CONFIGURATION

- Multidrop bus with maximum of 127 nodes. See Figure 8.1.
- Balanced differential cable, 120-ohm impedance, terminated.
- All nodes are transceivers.
- Many embedded controllers contain an integrated CAN interface.
- Asynchronous transmission.

LOGIC LEVELS

CAN uses unusual logic states on the bus. These are called recessive and dominant. Normally the bus levels rest at the recessive state that is

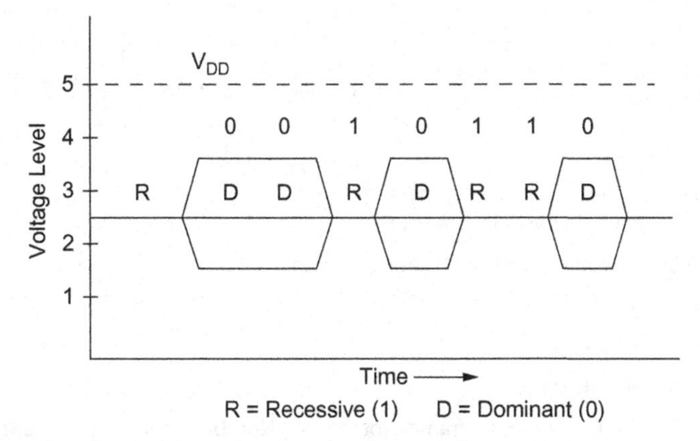

Figure 8.2 CAN bus voltages showing dominant and recessive states.

usually one half the supply voltage V_{DD} or V_{CC}. If V_{DD} is typically 5V, the recessive state is 2.5V. This is considered a binary 1 or high. The dominant state is when the bus difference voltage is > 1.5V. This is binary 0 or low. The bus lines are designated CANH (high) and CANL (low). Coding is NRZ. See Figure 8.2.

PROTOCOL

Figure 8.3 shows the simplified CAN frame with bits and bytes defined. Note: The arbitration field is the node's address or ID. It can be 11-bits in standard CAN or 29-bits in extended CAN. The arbitration field is also a priority designator with the lowest binary values having the highest priority.

- Transmission is asynchronous frames byte-by-byte with start and stop bits.
- Access method is carrier sense multiple access with collision detection (CSMA/CD) called bitwise arbitration (BA).
- Bus normally idles at the recessive (logic 1) state.
- To transmit, a node first listens for inactivity then sends a frame to the bus which all nodes see. Then a node decides whether it should respond to accept data or send data.
- If more than one node tries to transmit at a time, the node with the highest priority (lowest arbitration value) wins the contention. Other nodes stop transmitting and wait for an idle bus and then attempt transmission again.

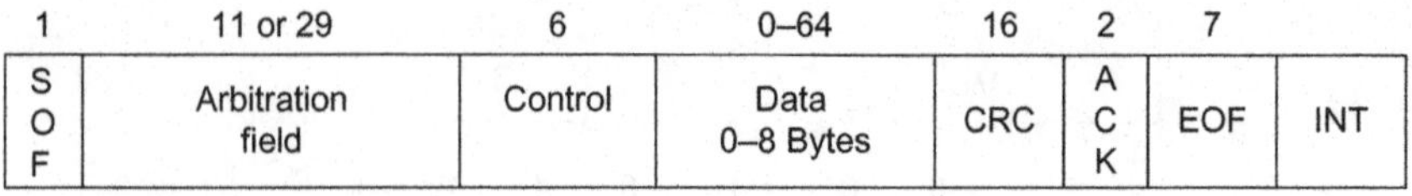

1	11 or 29	6	0–64	16	2	7	
SOF	Arbitration field	Control	Data 0–8 Bytes	CRC	ACK	EOF	INT

SOF – Start of frame

CONTROL – Identifier Extension (IDE)
and Data Length Code (DLE)

CRC – Cyclical Redundancy Check Error Detection

ACK – Acknowledge

EOF – End of Frame

INT – Intermission (Bus Free)

Figure 8.3 Simplified CAN protocol frame. Not to scale. Numbers indicate bits in each field.

PROTOCOL VARIATIONS

There are numerous variations of the CAN bus specifically CAN 2.0A (11-bit arbitration field) and CAN 2.0B (29-bit arbitration field). The other variations are primarily higher-level protocols that exist at OSI layers 3 through 7. All use standard CAN layers 1 and 2. Some of the most widely used are:

- CANopen – Used for non-automotive, mostly industrial applications (industrial machine control, medical, maritime, railway, etc.
- DeviceNet – Used mainly in factory automation. Maximum number of nodes is 64 and speeds are 125, 250, and 500 kb/s. Maximum range is 500 m at 125 kb/s.
- J1939 – Defined by the Society of Automotive Engineers (SAE) for truck and bus controls.

IC SOURCES

- Analog Devices
- Maxim Integrated
- Microchip Technology
- NXP
- ON Semiconductor
- ST Microelectronics
- Texas Instruments

CHAPTER NINE

DMX512

APPLICATIONS

- Stage lighting and effects via control boards.
- Control of dimmers, foggers, rotators, hazers, etc.

SOURCE

Original source United States Institute of Theatre Technology (USITT).

NATIONAL OR INTERNATIONAL STANDARD

- American National Standards Institute (ANSI)
- Entertainment Services and Technology Association (ESTA)
- Professional Lighting and Sound Association (PLASA)

KEY FEATURES

- A scalable network for controlling hundreds of electrical and electronic devices on a theater stage by way of a master control board.
- Based on the RS-485 bus standard. (See Chapter 28 on RS-485.)

DATA RATE

250 kb/s

CABLE MEDIUM

- Shielded twisted pair preferred, 120-ohm impedance.
- Unshielded CAT5 cable can be used if noise is not a problem.

Handbook of Serial Communications Interfaces.
Doi: http://dx.doi.org/10.1016/B978-0-12-800629-0.00009-7

- Microphone cable was originally used and can work over short distances but is not recommended.

CONNECTORS

- 5-pin XLR connector preferred. See Figure 9.1a.
- 3-pin XLR sometimes used. Not recommended because it is the same as an audio connector. Figure 9.1b.
- RJ-45 is the choice if CAT5 is used.

RANGE

Typically less than 500 m.

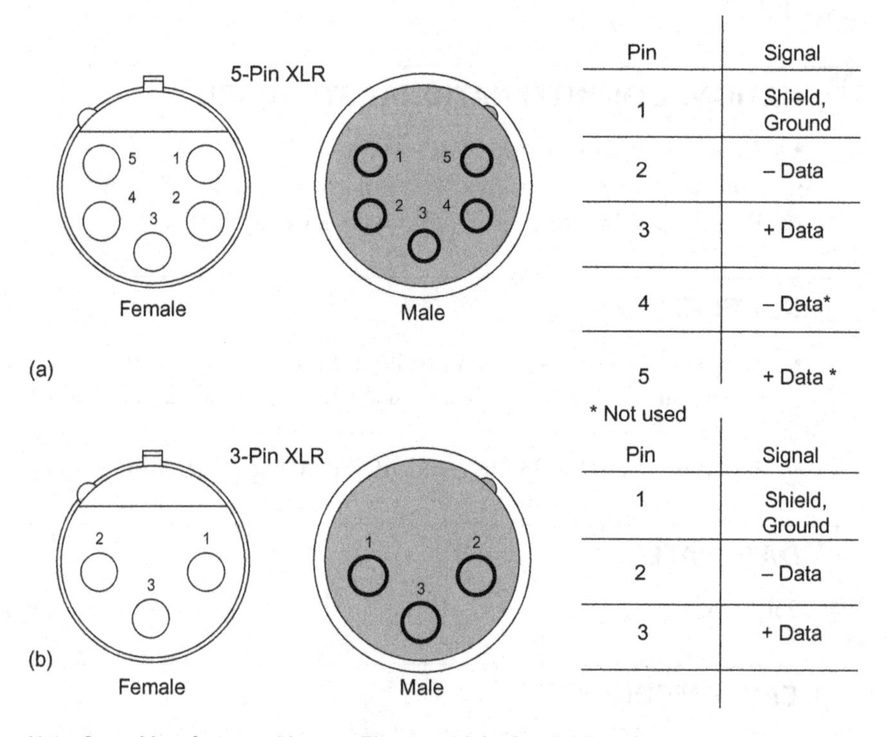

Pin	Signal
1	Shield, Ground
2	– Data
3	+ Data
4	– Data*
5	+ Data *

Pin	Signal
1	Shield, Ground
2	– Data
3	+ Data

Figure 9.1 Standard DMX512 connectors (a) recommended, (b) sometimes used.

NETWORK CONFIGURATION

- Multidrop bus with up to 32 nodes.
- Simplex operation only.
- Balanced signal connections.
- Devices may be daisy-chained. Most devices have an INPUT port and an OUTPUT or THRU port. See Figure 9.2.
- Up to 512 individual devices may be controlled through dimmer racks that serve up to 96 devices each.
- A 120-ohm termination must be used at the end of the chain.

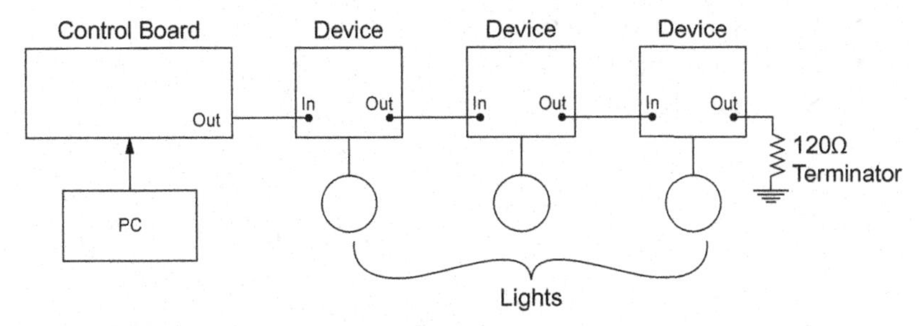

Figure 9.2 Standard bus/daisy-chain configuration.

LOGIC LEVELS

Binary 0 (SPACE): +2 to +6V, Binary 1 (MARK): −2 to −6V.

PROTOCOL

- Asynchronous frame with one byte of data, a start bit, and two stop bits. 11-bit word.
- Light level control is by data value: 0–255.
- A sequence of words called a packet is sent to control multiple lights.
- In idle mode, the line is high. A packet begins with a break, a low state that occurs for at least 88 μs.
- After the break, a mark after break (MAB) is transmitted. This is an 8 μs high pulse.
- Next is a start code (SC). This is a frame with a zero value to signal the beginning of a packet. A zero indicates dimmers.

- No address is transmitted. Each light is assigned an address (1 to 512) but that address is determined by the dimming code position in the packet. Frame 1 after the start code is for dimmer 1, frame 2 after the start code is for dimmer 2, and so on. The light dimmer circuits count the frames to determine if the code is for them. The sequence value after the start code determines the target dimmer.
- The packet ends with a break.

IC SOURCES

- Analog Devices
- Cypress Semiconductor
- Maxim Integrated
- See list for RS-485.

CHAPTER TEN

FlexRay

APPLICATIONS

Automotive connections between multiple electronic control units (ECUs) and various sensors and actuators.

SOURCE

FlexRay Consortium

NATIONAL OR INTERNATIONAL STANDARD

None

KEY FEATURES

- High data rates to 10 Mb/s.
- Time synchronized clocks.
- Deterministic data.
- Redundancy for reliability.

DATA RATE

1–10 Mb/s.

CABLE MEDIUM

Unshielded twisted pair, one or two pairs plus power and ground.

RANGE

Within a vehicle.

Handbook of Serial Communications Interfaces.
Doi: http://dx.doi.org/10.1016/B978-0-12-800629-0.00010-3

NETWORK CONFIGURATION

Supports multiple topologies such as multidrop bus, star and hybrid bus-star configurations.

LOGIC LEVELS

Binary 1: +5V, Binary 0: 0V. Nominal.

PROTOCOL

- Multiple configuration modes.
- Time division multiple access (TDMA) method provides fixed precise time segments to ensure deterministic data transmission if needed.
- Both static and dynamic frame segments for deterministic and non-deterministic data needs, respectively.
- Frame format consists of:
  - Header of five bytes that includes status, frame ID, payload length, header CRC, and cycle count bits.
  - Payload of up to 254 bytes.
  - Trailer of three 8-bit CRCs.

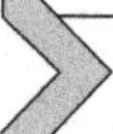

IC SOURCES

- Freescale
- NXP
- ON Semiconductor
- STMicroelectronics

CHAPTER ELEVEN

Foundation Fieldbus

APPLICATIONS

Industrial automation control in refineries, petrochemical plants, power plants, food and beverage plants, pharmaceutical factories, nuclear power plants, etc.

SOURCE

International Society of Automation (ISA)

NATIONAL OR INTERNATIONAL STANDARD

- Fieldbus Foundation
- ANSI/ISA 50.02
- International Electrotechnical Commission (IEC) 61158

KEY FEATURES

- Local area network for transducers, actuators and other field devices in a process control environment.
- Two versions: H1 for low speed and HSE (High-Speed Ethernet).
- Supplies DC power to the field devices.
- Low speed network can link to Ethernet.

DATA RATE

- H1: 31.5 kb/s
- HSE: 100/1000 Mb/s

Handbook of Serial Communications Interfaces.
Doi: http://dx.doi.org/10.1016/B978-0-12-800629-0.00011-5

CABLE MEDIUM

Unshielded or shielded twisted pair.

CONNECTORS

Standard screw terminal blocks.

RANGE

Up to 1900 m or 6232 feet. (Total of all wiring links.) Up to 9500 m with up to four repeater circuits.

NETWORK CONFIGURATION

- Multidrop bus. A trunk line with one or more spurs. See Figure 11.1.
- Multiple possible variations: point-to-point, daisy-chain, tree, star.
- Up to 32 field devices can be accommodated.
- Half duplex.
- Terminators consisting of a 100-ohm resistor and 1 μF capacitor in series must be used at the originating transmitter and at the device load.

LOGIC LEVELS

- DC voltage on the line, 9 to 32V.
- Logic signal is superimposed on the DC level, 0.75 to 1V peak-to-peak.
- Manchester Biphase-L encoding for embedded clock signal recovery.

PROTOCOL

- Layers 1, 2, and 7 of the OSI model.
- Proprietary: Fieldbus Foundation.

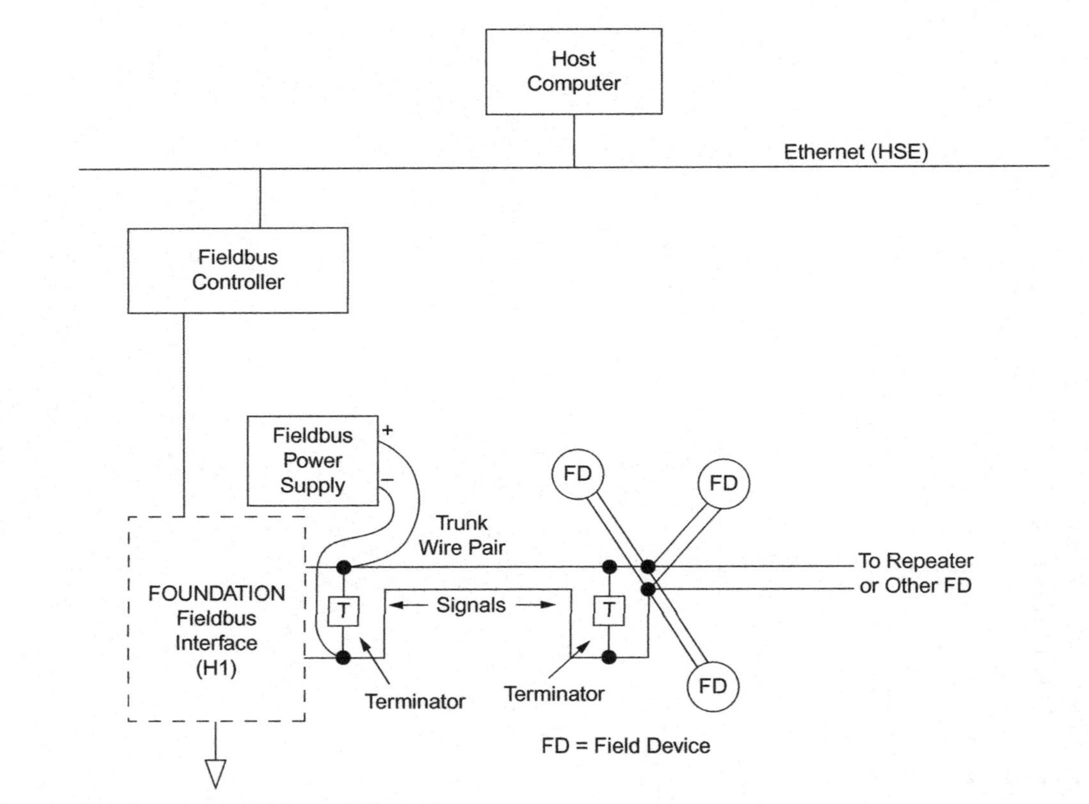

Figure 11.1 Standard H1 foundation fieldbus configuration.

IC SOURCES

- Aniotek
- Fuji
- ON Semiconductor
- SMAR Research
- Softing
- Yamaha Kagoshima

CHAPTER TWELVE

Highway Addressable Remote Transducer Protocol (HART)

APPLICATIONS

- Process control
- Factory automation
- Industrial control
- Pipelines

SOURCE

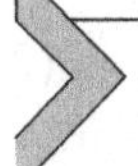

Rosemount developed HART and released it as an open standard that is managed by the HART Communications Foundation.

NATIONAL OR INTERNATIONAL STANDARD

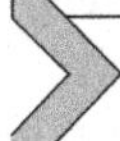

International Electrotechnical Commission (IEC) 61158

KEY FEATURES

- A digital networking technology for data transmission, provisioning, diagnostics, troubleshooting, and status of deployed transducers (sensors, actuators, etc.)
- Uses the existing 4–20 mA analog transducer control loop for communications.
- HART modulates the digital data (FSK) which is superimposed on the analog DC control voltage.
- Ideal for very long reaches between sensors and controllers.
- Excellent noise immunity.
- A wireless HART version available.

Handbook of Serial Communications Interfaces.
Doi: http://dx.doi.org/10.1016/B978-0-12-800629-0.00012-7

DATA RATE

1200 bps. 2 to 3 data updates (frames or packets) per second. Higher rate (3600 bps) uses C8PSK for a five-fold increase in update rate to 10 to 12 packets per second.

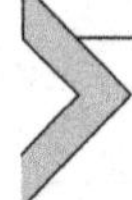

CABLE MEDIUM

Shielded twisted pair.

RANGE

Less than 10,000 feet. Depends on cable capacitance, resistance, and number of nodes on the network.

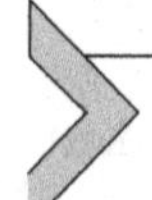

NETWORK CONFIGURATION

- Point-to-point using existing analog 4–20 mA control loop. See Figure 12.1. Most common configuration.
- Transmitter unit is usually a sensor that operates the current loop to convert the measurement value (temperature, pressure, flow, etc.) into a proportional current between 4 mA (minimum) and 20 mA (maximum). Current at the controller receiver

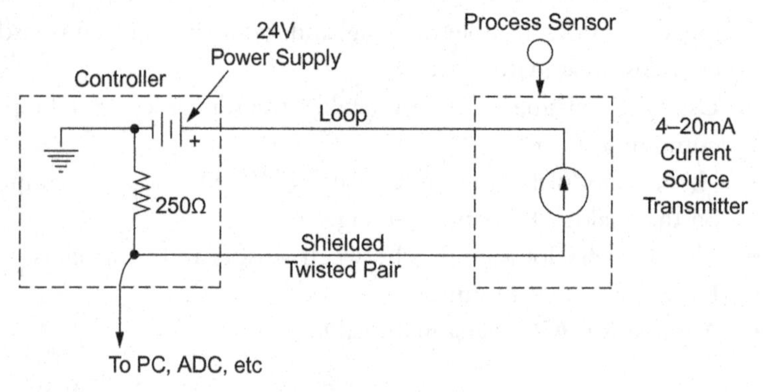

Figure 12.1 Basic 4–20 mA analog process loop.

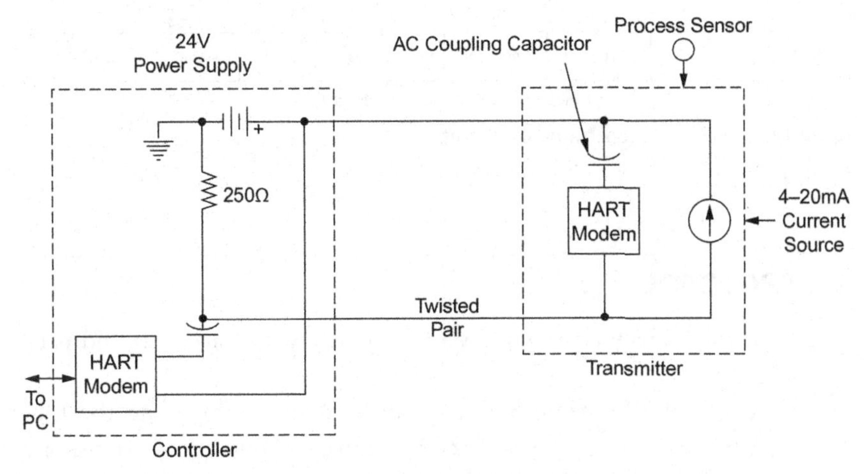

Figure 12.2 Simplified loop connection show HART modems.

is converted into a voltage across a 250-ohm load resistor. From there the proportional voltage (1 to 5V) is displayed, digitized in an ADC or otherwise processed and possibly used as feedback for closed loop control.

- HART circuitry is implemented in a single IC which includes the modem as well as related amplifiers and other circuits. The modem communicates with the sensor circuitry to perform data transmission, commissioning, status monitoring, troubleshooting, etc. See Figure 12.2. Note that modems are ac coupled.
- Multipoint can be implemented with two devices on a loop. Otherwise multiple devices can share a single loop if all measurement or control data is digital using HART. Control process variables are digitized and transmitted as data with HART. Loop current is fixed at 4 mA.

LOGIC LEVELS

Frequency shift keying (FSK) modulation of a phase continuous sine wave carrier by NRZ data with 1200 Hz as a binary 1 and 2400 Hz as a binary 0. DC control value is not affected since the sine wave averages to zero over time. The FSK signal is approximately ±500 mA peak-to-peak across the loop wires. An alternate modulation scheme uses C8PSK (continuous eight phase shift keying) to produce a higher data rate of 3600 bps.

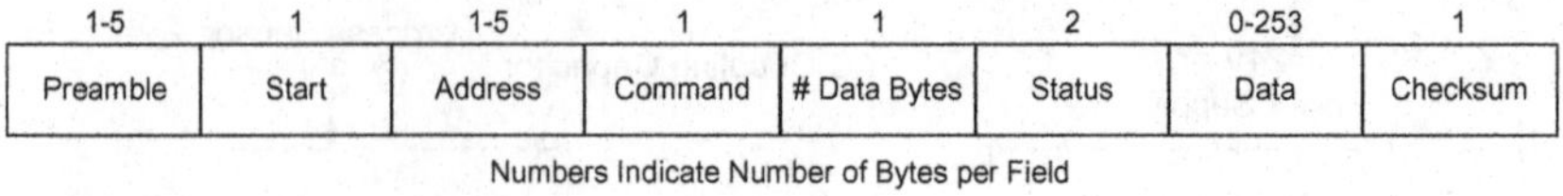

Preamble	Start	Address	Command	# Data Bytes	Status	Data	Checksum
1-5	1	1-5	1	1	2	0-253	1

Numbers Indicate Number of Bytes per Field

Figure 12.3 HART protocol frame or packet.

PROTOCOL

- 8-bit characters are sent asynchronously with start, stop, and parity bits.
- The HART protocol follows the OSI model by using the following layers: 1-Physical, 2-Data Link, 3-Network, 4-Transport, and 7-Application.
- Figure 12.3 shows the HART protocol frame.

IC SOURCES

- Analog Devices
- Maxim Integrated
- ON Semiconductor
- Texas Instruments

CHAPTER THIRTEEN

Inter-Integrated Circuit (I^2C) Bus

APPLICATIONS

- Interconnections between ICs on a printed circuit board (PCB).
- Short links between PCBs or other equipment.

SOURCE

NXP Semiconductor (formerly Philips Semiconductor).

NATIONAL OR INTERNATIONAL STANDARD

None

KEY FEATURES

- Low cost.
- Simple.
- Internal clock.
- Widely used.
- Most embedded microcontrollers include an I^2C interface.
- Many IC suppliers.

DATA RATE

100, 400 kb/s, 1, 3.4 Mb/s

CABLE MEDIUM

- Copper lines on a PCB.
- Four-wire shielded or unshielded cable.

Handbook of Serial Communications Interfaces.
Doi: http://dx.doi.org/10.1016/B978-0-12-800629-0.00013-9

RANGE

Depends on data rate and number of connected nodes. Cable capacitance limited (400 pF max.). Less than a foot is typical for PCBs. Usually no more than several feet. Up to 10 m at lowest data rate.

NETWORK CONFIGURATION

- Multidrop bus. See Figure 13.1.
- Unbalanced.
- Master-slave arrangement.
- Multiple masters can be accommodated but only one may use the bus at a time.
- Number of slave nodes depends on transmission medium capacitance usually no more than 20 or 30 slaves.
- Separate data (SDA) and clock (SCL) lines.
- Transmitters use open drain MOSFETs with external pull up resistors.

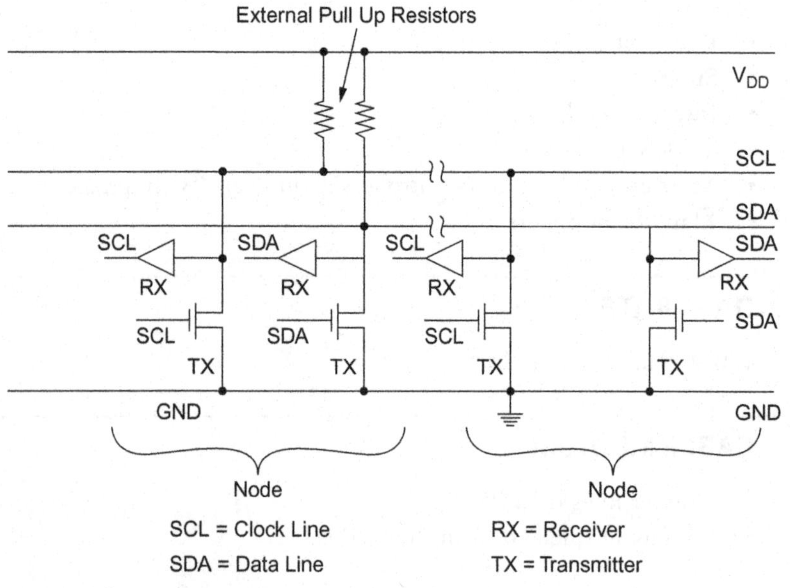

Figure 13.1 I^2C bus showing two nodes.

LOGIC LEVELS

Binary 0: zero or near zero. Binary 1: Depends on supply voltage usually 5V but also 3.3V and possibly others. Data format NRZ.

PROTOCOL

- Figure 13.2 shows the protocol frame.
- A Start bit begins a frame followed by a 7- or 10-bit slave address. Next is a read/write bit.
- An acknowledge (Ack) bit is sent by the slave if the address is recognized.
- Data is transmitted one byte at a time followed by an Ack bit.
- Frame ends with a stop bit.

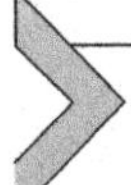

INTERFACE AND PROTOCOL VARIATIONS

- ACCESS.bus is a software protocol developed as an alternative to RS-232 to connect peripherals (keyboards, bar code readers, etc.) to computers. It uses the I^2C bus for the physical layer. This variation is no longer widely used.
- System Management Bus (SMBus) was defined by Intel for communications between chips and PCBs in computer motherboards. It is similar to I^2C but clock is in the 10–100 kb/s range while voltage and timing specifications are tighter. Protocol differences with I^2C include ACK/NACK, address resolution, time out, and packet error checking.
- Power Management Bus (PMBus) is a variation of the SMBus that is defined specifically for power supplies and power ICs. Clock rates up to 400 kb/s and data blocks to 255 bytes are allowed. Special power commands are defined in the protocol.

Number of Bits

1	7 or 10		1	8	1	8	1	8	1	1
Start	Address	R/W	ACK	Data 1	ACK	Data 2	ACK	Data N	ACK	Stop

Figure 13.2 Basic I^2C protocol frame.

- Adaptive Voltage Scaling Bus (AVSBus) is a complement to the PMBus. It provides a point-to-point link between a point of load (POL) voltage controller and a load. The prime application is to control processor voltages statically or dynamically for improved energy efficiency. It uses a two-wire interface (TWI) and protocol similar to PMBus.
- TWI is the same as I^2C interface. Atmel uses this name on its I^2C products to avoid patent/trademark conflicts with NXP (Philips).

IC SOURCES

- Analog Devices
- Atmel
- Cypress
- Freescale
- Infineon
- Intel
- Intersil
- Maxim Integrated
- Microchip Technologies
- NXP
- ON Semiconductor
- Silicon Laboratories
- STMicroelectronics
- Texas Instruments
- Xicor

CHAPTER FOURTEEN

IO-Link

APPLICATIONS

Industrial sensor and actuator connections.

SOURCE

IO-Link Consortium

NATIONAL OR INTERNATIONAL STANDARD

IEC 61131-9

KEY FEATURES

- Simple.
- Very low cost.
- Designed to work with a fieldbuses like Profibus or ProfiNet.
- Often used with programmable logic controllers (PLCs).

DATA RATE

4.8, 38.4, or 230.4 k baud.

CABLE MEDIUM

Three-wire or 5-wire standard industrial cable with power, ground, and data I/O connections. Connectors types M5, M8, and M12.

RANGE

20 m, maximum cable length

Handbook of Serial Communications Interfaces.
Doi: http://dx.doi.org/10.1016/B978-0-12-800629-0.00014-0

NETWORK CONFIGURATION

- Direct point-to-point connections between sensor/actuator and I/O ports on master controller.
- Byte-by-byte transfers using UART asynchronous format with start, stop, and parity (even) bits.

LOGIC LEVELS

Binary 0: 24V, Binary 1: 0V typical. NRZ encoding.

PROTOCOL

- Master initiates all communications.
- Master interrogates desired sensor/actuator.
- Three types of data are exchanged: cyclic or process data, acyclic or device service data and status, and sensor/actuator events.
- Data is exchanged in frames called telegrams.
- A master request telegram is two bytes long with a CMD byte and a CHK/TYPE byte. The CMD byte contains read/write indicator bit, a 2-bit data channel, and a 5-bit address. The CHK/STAT byte contains a 2-bit data channel and a 6-bit checksum.
- The device reply telegram is either two or three bytes long. The two-byte telegram contains one byte of process data and a CHK/STAT byte. The three-byte telegram contains one byte or service data, one byte of process data, and a CHK/STAT byte.

IC SOURCES

- Freescale Semiconductor
- Infineon Technologies
- Linear Technology
- Maxim Integrated
- Texas Instruments

CHAPTER FIFTEEN

Inter-IC Sound (I^2S) Bus

APPLICATIONS

Interconnection of digital sound ICs and other digital audio devices.

SOURCE

NXP Semiconductor (formerly Philips Semiconductor)

NATIONAL OR INTERNATIONAL STANDARD

None

KEY FEATURES

- Used with audio A/D and D/A converters (codecs), DSP, filters, compact disk processors, etc.
- A separate clock line shared by transmitter and receiver that eliminates timing errors and jitter.
- Facilitates digital (PCM) stereo audio transmission.
- Digital word size adjustable up to 64-bits (16, 24, 32, 64-bits).
- Works with standard digital audio sampling rates of 32, 44.1, 48, 96, 192 kb/s, and others.

DATA RATE

2.5 MHz typical. Up to 12.288 MHz.

CABLE MEDIUM

Short copper connections on a PCB less than one foot. Short non-defined wire cable, several feet maximum.

Handbook of Serial Communications Interfaces.
Doi: http://dx.doi.org/10.1016/B978-0-12-800629-0.00015-2

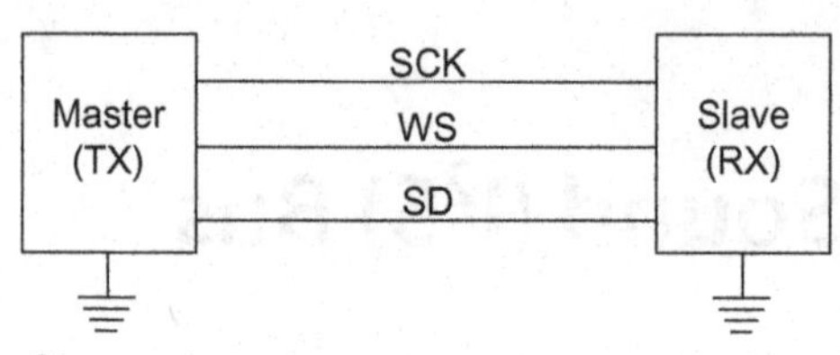

Figure 15.1 Typical I²S bus connections.

RANGE

Typically less than one foot or several feet maximum.

NETWORK CONFIGURATION

- Master-slave connection with the transmitter as master and receiver as slave. See Figure 15.1.
- Can be configured with receiver as master or a separate controller as master and other devices as slaves.
- Serial clock (SCK) line, word select (WS), and serial data (SD) lines with common ground.

LOGIC LEVELS

Standard TTL levels. Binary 0: <0.8V, Binary 1: >2V.

PROTOCOL

- Two independent stereo digital audio signals are time multiplexed on the serial data channel.
- Word select (W) line designates which stereo channel (left or right) is on the data line. Left channel transmits first.
- Digital data is usually transmitted MSB first.

IC SOURCES

- Analog Devices
- Cirrus Logic
- Cypress Semiconductor
- NXP
- Texas Instruments

Local Interconnection Network (LIN)

APPLICATIONS

An automotive network for communications with simple sensors and actuators that use minimal data. Typical applications include AC systems, doors, windows, seats, steering column, climate controls, switches, wipers, sunroof, etc.

SOURCE

LIN Consortium

NATIONAL OR INTERNATIONAL STANDARD

ISO 9141

KEY FEATURES

- Simple architecture.
- Very low cost.
- Typically used as a sub-bus to CAN automotive systems.
- Minimal EMI.
- Guaranteed latency.

DATA RATE

Commonly 19.2 kb/s but also supports 2400 and 9600 b/s.

CABLE MEDIUM

Single wire plus automotive ground return.

Handbook of Serial Communications Interfaces.
Doi: http://dx.doi.org/10.1016/B978-0-12-800629-0.00016-4

RANGE

40 m maximum at 19.2 kb/s.

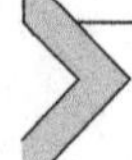

NETWORK CONFIGURATION

- Multidrop bus. See Figure 16.1.
- Master-slave configuration with one master and up to 16 slaves.
- Open drain/collector with external pull up resistors.
- Each node has an appropriately programmed embedded microcontroller.
- Self-synchronizing based on master.
- Checksum error detection.

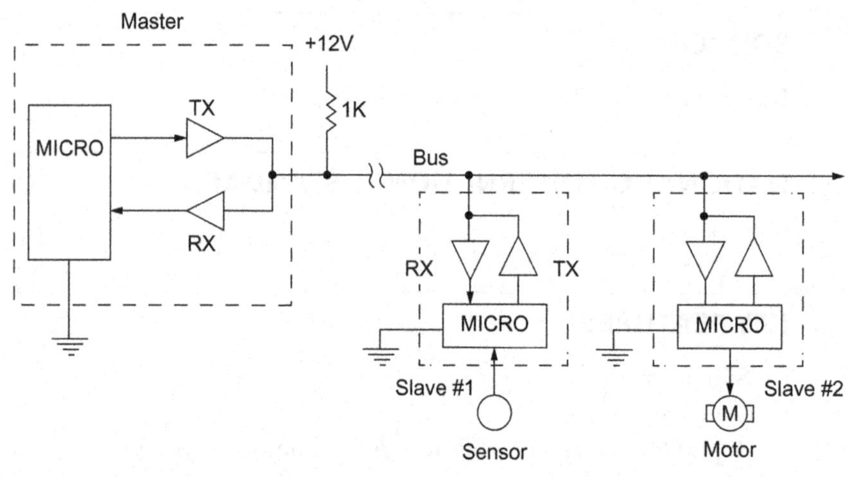

Figure 16.1 Simplified LIN bus configuration.

LOGIC LEVELS

Binary 0: 0V, Binary 1: 12V. Levels based on vehicle battery voltage, nominally 12V. NRZ coding.

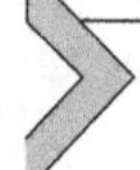

PROTOCOL

- Master controls all communications, sequence, priority, and order.
- No bus arbitration or access method used.

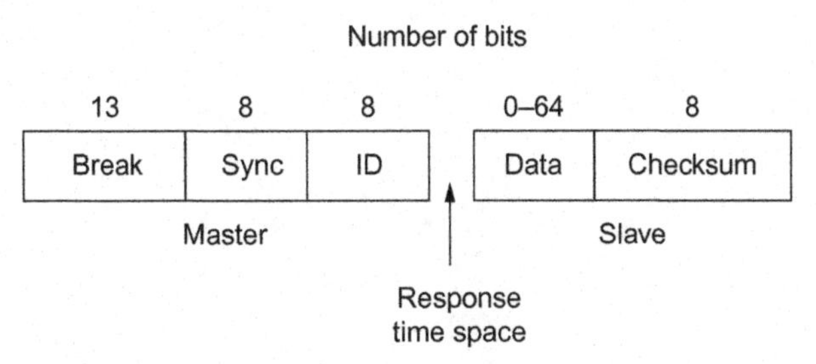

Figure 16.2 LIN protocol frame.

- Uses UART serial data format with start and stop pulses and 8-bit data fields.
- Master initiates communication by sending a break field, a sync field, and a slave identifier (ID). See Figure 16.2.
- The break field is a 13-bit signal to the slaves that a new frame is beginning.
- Sync field is hex 55 and used by the slave to determine the master clock rate and initiate synchronization.
- ID field is a 6-bit address plus two parity bits.
- The addressed slave responds with a data field of zero to eight bytes and an 8-bit checksum.
- Checksum may include data or data plus ID.

IC SOURCES

- Atmel
- Freescale
- Fujitsu
- Infineon
- Melexis
- Maxim Integrated
- Microchip
- NEC
- NXP
- On Semiconductor
- Texas Instruments
- ZMD

CHAPTER SEVENTEEN

Meter Bus (M-Bus)

APPLICATIONS

Remote reading of utility meters (water, gas, electric). Alarms, lighting, and heating control.

SOURCE

Developed by Professor Horst Ziegler of the University of Paderborn with Texas Instruments Deutchland GmbH and Techem GmbH.

NATIONAL OR INTERNATIONAL STANDARD

European standard: EN 13757-2, EN 13757-3

KEY FEATURES

- Large number of potential nodes.
- Long range connections.
- Low cost.
- Low power consumption.
- Simplicity.
- Robust and reliable.
- Slaves powered by master.

DATA RATE

300 to 9600 b/s. 2400 b/s default. 19.2, 38.4, and 115.2 kb/s are possible on shorter cable runs.

Handbook of Serial Communications Interfaces.
Doi: http://dx.doi.org/10.1016/B978-0-12-800629-0.00017-6

CABLE MEDIUM

Two-wire unshielded twisted pair telephone cable.

RANGE

1000 m. Maximum distance between slave and a repeater is 350 m.

NETWORK CONFIGURATION

- Single master and up to several hundred (250) slaves (utility meters).
- Two-wire multidrop bus.
- Repeaters can be used to extend the range.

LOGIC LEVELS

Master to slave: Binary 1 (mark): 36V, Binary 0 (space): 24V.
Slave to master: Binary 1 (mark): <1.5 mA, Binary 0 (space): 11–20 mA.

PROTOCOL

- Uses standard asynchronous UART single byte transfer with start, even parity and stop bits.
- Uses standard OSI model layers 1, 2, 3, and 7.
- Half duplex, no access method as master only communicates with one slave at a time.
- Communication is by way of three types of "telegrams": Single Character, Short Telegram, and Long Telegram.
- Single Character telegram (hex E5) acknowledges that a telegram has been received.
- Short Telegram has the start character hex 10 and consists of five characters to tell a slave to send data to the master. It is made up of one character C and A fields and a checksum.
- Long Telegram uses a variable number of characters with start character hex 68. It is used by the master to send commands to a slave and used by a slave to send meter data to the master. It is made up of one character C, A and CI fields and a checksum.

- The C, A, CI, and L fields are used to address a slave, send commands, set data rate, state number of data bytes, and other information.
- Data can be up to 246 bytes.

IC SOURCES

- ON Semiconductor
- Texas Instruments

CHAPTER EIGHTEEN

Microwire

APPLICATIONS

- Chip-to-chip connections on a printed circuit board (PCB) or between PCBs.
- Microcontroller to peripheral devices like memory chips, ADC, DAC, displays, sensors, etc.

SOURCE

Originated with National Semiconductor now part of Texas Instruments.

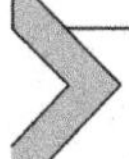

NATIONAL OR INTERNATIONAL STANDARD

None

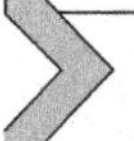

KEY FEATURES

- Older predecessor to the serial peripheral interface (SPI).
- Similar to the serial peripheral interface (SPI). See Chapter 35 on SPI.
- A subset of SPI, usually half duplex only.
- A later version called Microwire/Plus offers full duplex operation.
- Synchronous data transfer with single clock.
- Three-wire bus with input, output, and clock lines plus slave select lines.

Handbook of Serial Communications Interfaces.
Doi: http://dx.doi.org/10.1016/B978-0-12-800629-0.00018-8

DATA RATE

Unspecified. Often less than 250 kb/s. Zero up to about 3 Mb/s.

CABLE MEDIUM

Copper pattern on PCB or short ribbon cable between PCBs.

RANGE

Up to 1 m but typically less than one foot.

NETWORK CONFIGURATION

- Master-slave connections where the master is usually an embedded controller and slaves are peripheral chips. See Figure 18.1.
- Three lines: Master serial out slave in (SO), master serial in slave out (SI), and clock (SK).

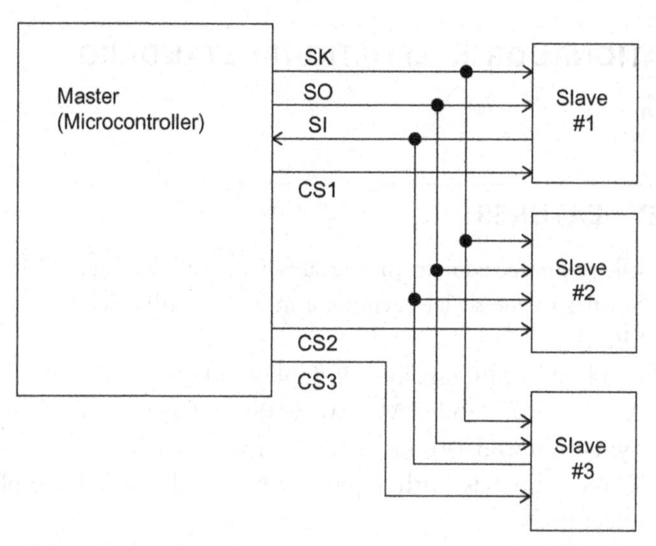

Figure 18.1 Microwire bus.

- Slaves must have a chip select (CSn) line usually implemented with a GPIO bus line from the master microcontroller.
- Slaves may be daisy-chained and a single common slave select line is used.

LOGIC LEVELS

TTL or CMOS compatible: Binary 0: <0.4V, Binary 1: >2.4V.

PROTOCOL

- Protocol usually proprietary to the specific application or operation of the slave ICs.
- Word size is typically 8-bits, but 12, 16, and other bit word sizes can be accommodated.
- Data transmission sequence:
- Master transmits a binary 0 to the selected slave.
- Slave sends data to master or master sends data to slave. MSB first.
- Different modes of operation specify whether data is transmitted or captured by the leading or trailing edge of the clock.

IC SOURCES

- Analog Devices
- Atmel
- Freescale
- Maxim Integrated
- Microchip Technology
- Texas Instruments

CHAPTER NINETEEN

Musical Instrument Digital Interface (MIDI)

APPLICATIONS

Connections and communications between computers, keyboards, and other musical instruments for command and control purposes.

SOURCE

MIDI Manufacturers Association (MMA) – 1983

NATIONAL OR INTERNATIONAL STANDARD

MIDI Manufacturers Association (MMA)

KEY FEATURES

- A serial interface that carries commands from a computer, keyboard, or MIDI controller to send commands and messages to other instruments.
- MIDI does not carry digitized audio.
- Allows music to be stored as a series of commands and transmitted to two or more instruments for simultaneous playback.

DATA RATE

31.25 kbaud or 31.25 kb/s, 32 μs bit time.

CABLE MEDIUM

Shielded twisted pair. Connector is a 5-pin female DIN. See Figure 19.1. MIDI OUT is pin 5, MIDI IN is pin 4, cable shield is pin 2

Handbook of Serial Communications Interfaces.
Doi: http://dx.doi.org/10.1016/B978-0-12-800629-0.00019-X

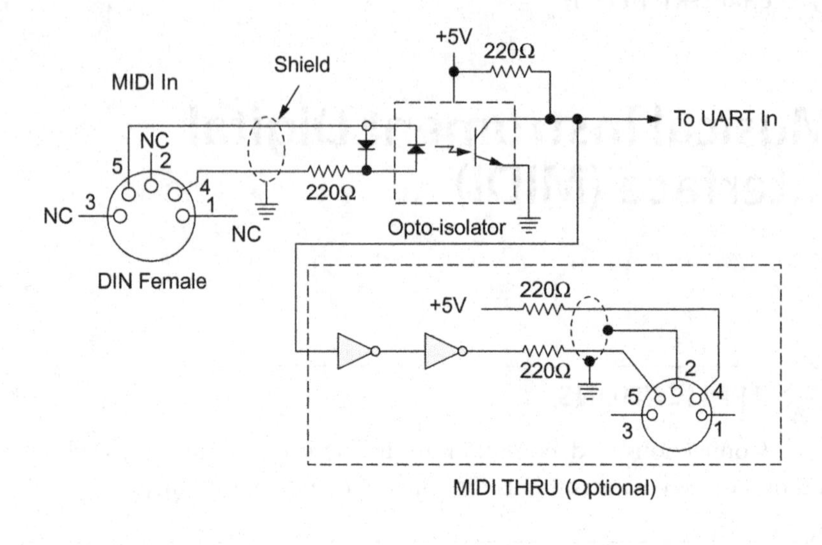

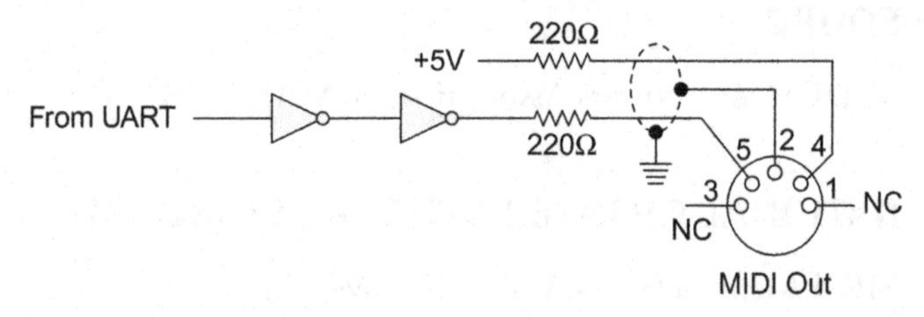

Figure 19.1 MIDI interface connections.

(OUT only). Pins 1 and 3 are not used. Separate jacks are used for input and output.

RANGE

Maximum recommended cable length is 50 feet.

NETWORK CONFIGURATION

Point-to-point or daisy-chained. Daisy-chained configuration makes use of a MIDI THRU connection that passes input to the output pin so multiple devices may be linked. See Figure 19.1.

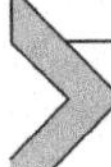

LOGIC LEVELS

TTL compatible. +5V supply. Binary 0 is current "on" (1.5 to 5 mA). Binary 1 is current "off." MIDI input is opto-isolated to minimize noise, feedback and ground loops. See Figure 19.1.

NOTE: While most equipment still uses the cabling and connectors described above, newer equipment uses the universal serial bus (USB) in place of or in addition to the traditional connections. The Firewire and Ethernet interfaces have also been used.

PROTOCOL

- Asynchronous format using standard UART with 8-bit words and start and stop bits.
- Communications is by one or more messages.
- Messages consist of sequence of command and data bytes that specify operations.
- Command bytes have the MSB as binary 1 with the range of hex 80–FF. Command bytes tell the instrument what to do such as play a note.
- Data bytes have the MSB as binary 0 with the range of hex 00–7F. Data bytes state what note to play and how to play it.

IC SOURCES

Multiple standard compatible microcontroller, UART, logic, and interface sources.

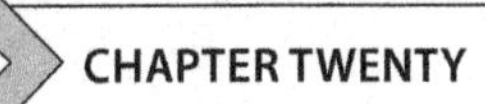

CHAPTER TWENTY

MIL-STD-1553

APPLICATIONS

- Military avionics.
- Spacecraft on-board data handling.

SOURCE

U.S. Department of Defense (DoD)

NATIONAL OR INTERNATIONAL STANDARD

Society of Automotive Engineers (Aerospace Division)

KEY FEATURES

- Highly reliable dual redundant interface for multiple sensors and actuators.
- Low bit error rate.

DATA RATE

1 or 10 Mb/s.

CABLE MEDIUM

Shielded twisted pair, nominal impedance between 70 and 85 ohms. Redundant pairs frequently used.

Handbook of Serial Communications Interfaces.
Doi: http://dx.doi.org/10.1016/B978-0-12-800629-0.00020-6

RANGE

Generally unspecified. Depends on cable type and issues related to signal attenuation or delay time. Stub length for connected terminals should be less than 20 feet.

NETWORK CONFIGURATION

- Differential data bus. See Figure 20.1
- Bus Controller (BC) with up to 31 Remote Terminals (RT).
- Transformer coupling to bus. See detail in Figure 20.2.
- Half duplex operation.
- Time division multiplex.

LOGIC LEVELS

Typical transmitter output level is 18–27V peak-to-peak. Manchester II bi-phase encoding for reliable clock recovery and no DC on the bus.

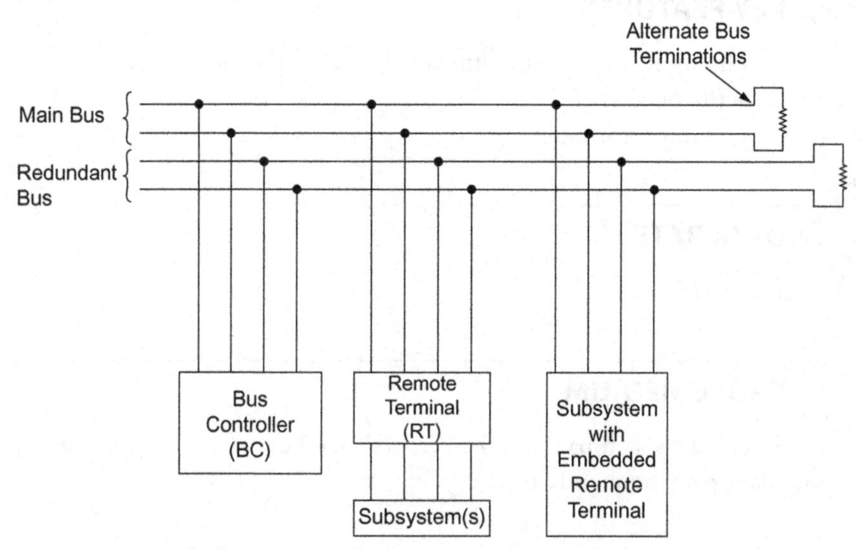

Figure 20.1 MIL-STD-1553 bus configuration. (Transformer coupling not shown.)

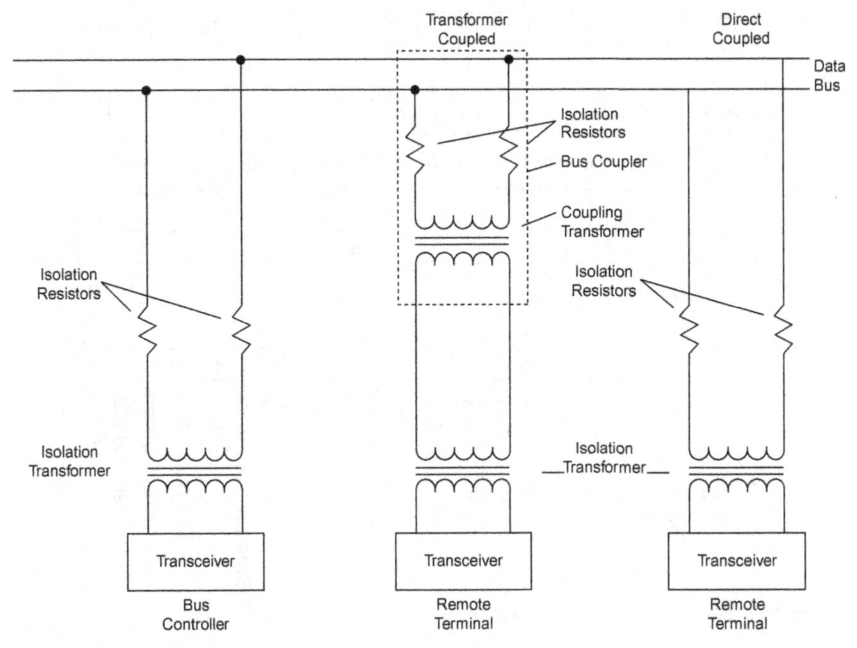

Figure 20.2 Bus transformer coupling.

PROTOCOL

- Bus controller (BC) initiates all operations to Remote Terminals (RT). Command/response.
- Allowed communications: BC to RT, RT to BC, RT to RT, broadcast, system mode control.
- Three word types: command, data, status. See Figure 20.3.
- 20-bit words, 3 sync bits, 16 data bits, one parity bit.
- Asynchronous operation with 4 μs gap between words or messages.
- Messages are one or more words.

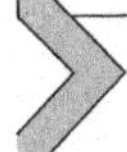

IC SOURCES

- Aeroflex
- C-MAC MicroTechnology
- Data Device Corp.
- Holt Integrated Circuits
- National Hybrid Inc.

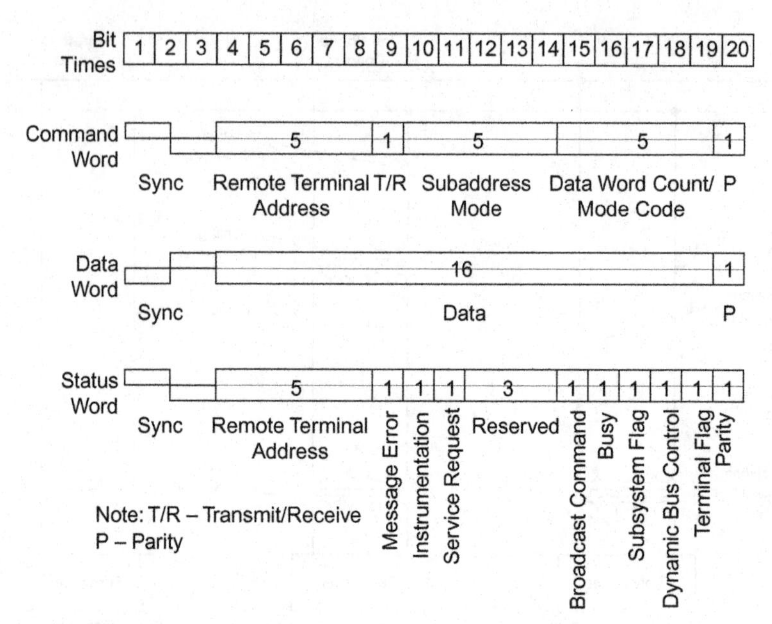

Figure 20.3 MIL-STD-1553 word formats.

CHAPTER TWENTY-ONE

Modbus

APPLICATIONS

Industrial monitoring and control especially with programmable logic controllers (PLCs).

SOURCE

Original source Modicon which was acquired by Schneider Electric.

NATIONAL OR INTERNATIONAL STANDARD

Modbus Organization.

KEY FEATURES

- Open source, royalty free protocol.
- Uses standard RS-485 or RS-232 physical interface.
- Simple implementation.
- Very widely used and supported.

DATA RATE

9600 b/s or 19.2 kb/s standard with 19.2 kb/s the default. Can also use slower standard modem rates (2400, 4800 b/s, etc.) and up to 115.2 kb/s.

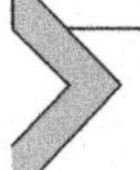

CABLE MEDIUM

Shielded or unshielded twisted pair.

Handbook of Serial Communications Interfaces.
Doi: http://dx.doi.org/10.1016/B978-0-12-800629-0.00021-8

CONNECTORS

Screw terminals, 9-pin D-shell, and RJ-45 connectors have been used.

RANGE

1000 m maximum for 9600 b/s, shorter for higher speeds. Tap lengths must be less than 20 m.

NETWORK CONFIGURATION

- Point-to-point or multidrop bus connections through taps or by daisy-chain.
- Up to 247 devices may be identified.
- Both ends of the bus must be terminated in 150 ohms.
- Master/slave relationship where only the master controls operations.

LOGIC LEVELS

Binary 0 (SPACE): +2 to +6V, Binary 1 (MARK): −2 to −6V.

PROTOCOL

- OSI model layers 1, 2, and 7 only.
- Two operational modes: remote terminal unit (RTU) and ASCII, RTU the most common.
- Data transferred in bytes using 11-bit asynchronous frame with start bit, data byte, even parity, and one stop bit.
- RTU frame format is shown in Figure 21.1.

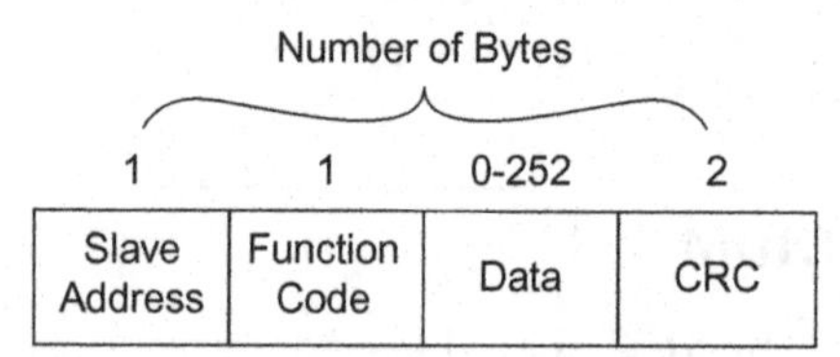

Figure 21.1 MODBUS RTU frame format.

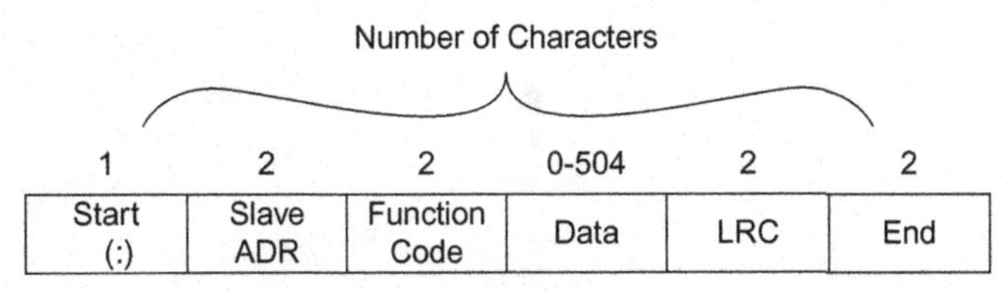

Figure 21.2 MODBUS ASCII mode frame format.

- Frames are separated by a time equal to 3.5 characters or 28-bits.
- In ASCII mode, each byte is sent two ASCII characters in a 10-bit format (1 start bit, 8-bits of data, 1 parity bit plus 1 stop bit or no parity and 2 stop bits).
- ASCII mode frame shown in Figure 21.2. Start character is a colon (:). LRC is a longitudinal redundancy check error detection code, END character is ASCII carriage return (CR) and line feed (LF) codes.

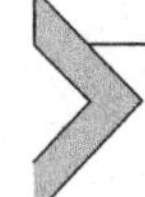

IC SOURCES

See Chapter 28 on RS-485.

CHAPTER TWENTY-TWO

On-Board Diagnostics (OBD) II

APPLICATIONS

Monitor the status of the various subsystems in a car or truck for diagnostic or performance purposes.

SOURCE

Required by the Environmental Protection Agency in U.S. vehicles since 1996 initially for pollution control testing.

NATIONAL OR INTERNATIONAL STANDARD

- Society of Automotive Engineers (SAE) J1962, J1850, and others.
- International Standards Organization (ISO) ISO 9141, ISO 15765, and others.

KEY FEATURES

- A standardized serial port for extracting data from vehicle sensors and subsystems for diagnosing vehicle problems and performance.
- Insurance companies use the port to obtain vehicle operations data for setting rates.
- Fleet driver monitoring operations.
- Multiple interface standards are defined by the vehicle manufacturers.
- The common standard is the OBD II 16-pin connector that is usually located under the dashboard within 2 to 3 feet of the steering column. Most ports have a cover to protect the connector pins. See Figure 22.1.

Handbook of Serial Communications Interfaces.
Doi: http://dx.doi.org/10.1016/B978-0-12-800629-0.00022-X

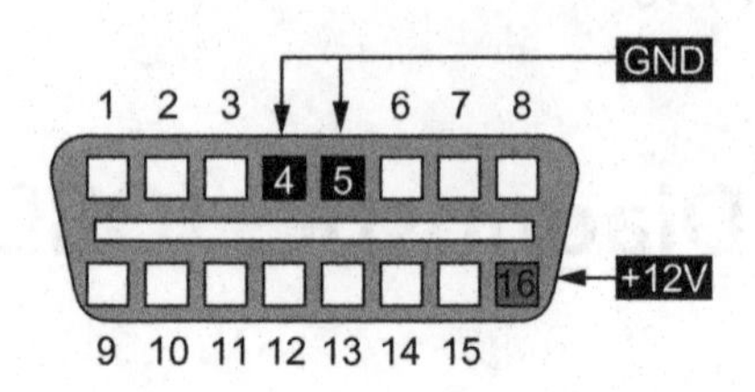

Pin	Function
1	—
2	J1850 Bus +
3	—
4	Chassis Ground
5	Signal Ground
6	CAN High
7	ISO 9141-2 K-line
8	—

Pin	Function
9	—
10	J1850 Bus –
11	—
12	—
13	—
14	CAN Low
15	ISO 9141-2 L-line
16	+12V DC

Figure 22.1 OBD II connector.

DATA RATE

Varies with the interface standard used. Low rates of 10.4 and 41.6 kb/s are the most common. CAN versions support 250 and 500 kb/s.

CABLE MEDIUM

Multiwire cable.

RANGE

Several feet.

NETWORK CONFIGURATION

- Point-to-multipoint. A master data collection device and multiple nodes monitoring vehicle sensors and subsystems.
- Typical collection devices include hand-held scanners, personal computers, data loggers, emission testing devices, or vehicle telematics (wireless: Bluetooth or Wi-Fi).

LOGIC LEVELS

Depends on specific standard. Examples:

- SAE J1850 Pulse width modulation (PWM)
  - Two-wire differential (pins 2 and 10)
  - Binary 1: +3.8–5V, Binary 0: <1.2V.
  - Binary 1: 8 µs, Binary 0: 16 µs
  - Data rate 41.6 kb/s (24 µs period)
- SAE J1850 Variable pulse width (VPW).
  - Single-ended signal on pin 2.
  - Binary 1: +7V, Binary 0: <1.5V
  - Binary 1: Low for 128 µs or high for 64 µs, Binary 0: Low for 64 µs or high for 128 µs.
  - Data rates 10.4 kb/s (96 µs period) or 41.6 kb/s (24 µs period)
- ISO 9141-2, ISO 14230, and KWP2000
  - Single-ended signal on pin 7.
  - Binary 1: +12V, binary 0: <2.4V.
  - 10.4 kb/s data rate, UART format.
- ISO 15765 CAN bus
  - Differential CAN signals on pins 6 (CANH) and 14 (CANL).
  - Binary 1: 3.5V (CANH), Binary 0: 1.5V (CANL).
  - Data rate 250 or 500 kb/s.

PROTOCOL

Multiple protocols unique to the manufacturer. Most use UART data formats and specific frame structures.

IC SOURCES

- Elm Electronics (ELM320, ELM322, ELM323, ELM325, ELM327, ELM328, ELM329 chips)
- OBD Solutions (STN1110 chip)

CHAPTER TWENTY-THREE

Power Management Bus (PMBus)

APPLICATIONS

Used for control and protection of power supplies, power monitor and control ICs, temperature monitors, battery chargers, and similar devices.

SOURCE

PMBus Implementers Forum

NATIONAL OR INTERNATIONAL STANDARD

System Management Interface Forum (SM-IF)

KEY FEATURES

- An expanded version of the System Management Bus. See related Chapter 29.
- Derivative of and mostly compliant with I^2C bus specifications.
- Works with all types of power supplies and power converters.
- Allows fine-tuning of voltage control, current levels, temperature settings, etc.
- Includes a dedicated Adaptive Voltage Scaling (AVS) bus to control processor voltages.

DATA RATE

10 to 400 kb/s.

CABLE MEDIUM

Copper connections on a printed circuit board.

Handbook of Serial Communications Interfaces.
Doi: http://dx.doi.org/10.1016/B978-0-12-800629-0.00023-1

RANGE

Less than 1 m.

NETWORK CONFIGURATION

Same as I²C with added interrupt and on–off control lines. Refer to Figure 23.1.

LOGIC LEVELS

Binary 0: <0.8V, Binary 1: >2.1V.

PROTOCOL

- Same as SMBus with some exceptions including the following. See Figure 23.1.
- Separate SMBALERT# interrupt line from SMBus added.

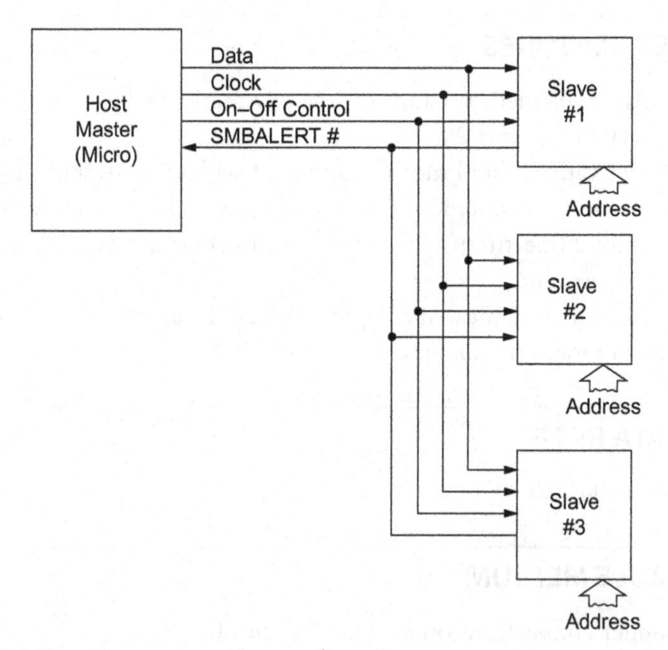

Figure 23.1 Power management bus configuration.

- Larger blocks of data up to 255 bytes may be sent.
- Additional command byte added for 256 more standard and manufacturer-specific operations.
- Separate physical control line between master and slaves is added to implement on–off control functions.
- Separate physical address assignments needed for slaves.

IC SOURCES

- Advanced Micro Devices
- Altera
- Analog Devices
- Atmel
- Cypress
- Fairchild
- International Rectifier
- Intel
- Intersil
- Linear Technology
- Maxim Integrated
- Microchip Technology
- NXP
- Semtech
- ST Microelectronics
- Texas Instruments
- Xilinx

CHAPTER TWENTY-FOUR

PROFIBUS

APPLICATIONS

Industrial monitoring and control, process automation, factory automation, motion control.

SOURCE

German industry and government. Siemens.

NATIONAL OR INTERNATIONAL STANDARD

- International Electrotechnical Commission (IEC) 61158, 61784
- DIN 19245
- EN 50170
- PROFIBUS User Organization
- PROFIBUS & PROFINET International

KEY FEATURES

- Multiple versions of target-specific applications in industry.
  - PROFIBUS Fieldbus Message Specification (FMS) – For Programmable Logic Controllers (PLCs). Non-deterministic.
  - PROFIBUS Decentralized Peripherals (DP) – Most common version. Factory automation. Deterministic.
  - PROFIBUS Process Automation (PA) – Monitor/control process control.
  - Manchester Bus Powered (MBP) – DC power and Manchester encoding over the data bus for the PA version. Similar to Foundation Fieldbus H1.
- Use of RS-485 for physical layer for FMS and DP versions.
- Open standard.

Handbook of Serial Communications Interfaces.
Doi: http://dx.doi.org/10.1016/B978-0-12-800629-0.00024-3

- Widely supported by multiple manufacturers.
- PROFINET standard allows PROFIBUS to be carried over industrial Ethernet.

DATA RATE

9.6 kb/s to 12 Mb/s. Determined by cable length. 31.25 kb/s for PA version.

CABLE MEDIUM

Shielded or unshielded twisted pair. 150 ohm. Optical fiber is an option.

RANGE

100 to 1200 m. Up to 1900 m for PA version. Up to 15 km with optical fiber.

NETWORK CONFIGURATION

- Multidrop bus.
- Star and ring configurations possible with optical fiber.
- Master/slave configuration.
- Half duplex. Slave polling.
- Token passing where only the node with the token can communicate.
- Up to 126 connected devices with 32 maximum per segment. Repeaters allowed.

LOGIC LEVELS

Binary 0 (SPACE): +2 to +6V, Binary 1 (MARK): −2 to −6V on PROFIBUS-DP with RS-485.

PROTOCOL

- OSI model layers 1, 2, and 7.
- Data sent in asynchronous bytes with start, stop, and parity bits.

- Basic protocol frame is called a telegram. Types of telegrams.
  - No data, 6 control bytes.
  - Fixed data (8-bytes), 6 control bytes.
  - Variable data (0–244 bytes), 9 to 11 control bytes.
  - Acknowledge or recognition, one byte.
  - Token, 3 bytes.
- Master telegrams start with a SYN pause of minimum 33 bits of binary 1.
- Figure 24.1 shows a typical data telegram from master.

SYN	SD2	LE	LE_r	SD2	DA	SA	FC	DU	FCS	ED

SYN	33 Bits, Logic 1
SD2	Start Delimiter 2
LE	Length of Data Units
LE_r	Length Repeated
DA	Destination Address
SA	Source Address
FC	Function Code
DU	Data Unit
FCS	Frame Check Sequence
ED	End Delimiter

Figure 24.1 Typical PROFIBUS telegram from master to slave, variable data.

IC SOURCES

- Altera
- Deutschmann Automation
- Freescale
- Hilcher
- HMS Anybus
- Innovasic
- KW-Software
- Molex
- Siemens
- Unigate

CHAPTER TWENTY-FIVE

RS-232

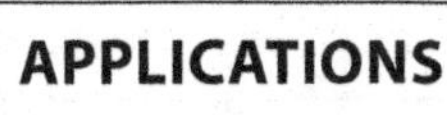

APPLICATIONS

- Originally used to connect teletypewriters and video computer terminals to modems.
- Once widely used to connect personal computers to peripherals like printers. Generally known as a "serial port" or serial communications interface (SCI).
- Still incorporated in many industrial automation devices, machine tools, programmable logic controllers (PLC), test instruments, embedded microcontrollers, and scientific equipment.

SOURCE

Electronics Industry Alliance (EIA) in 1962.

NATIONAL OR INTERNATIONAL STANDARD

American National Standards Institute (ANSI), Electronics Industry Alliance/Telecommunications Industry Association, EIA/TIA-232-F.

KEY FEATURES

- One of the oldest if not the oldest serial interface.
- Generally replaced by USB in personal computers and peripheral equipment.
- Still popular in industrial equipment.
- USB to RS-232 converters commonly available.
- Simple and reliable interface in a noisy environment.
- Low but adequate data rate.

Handbook of Serial Communications Interfaces.
Doi: http://dx.doi.org/10.1016/B978-0-12-800629-0.00025-5

DATA RATE

Standard states upper limit is 20 kb/s. Actual upper rate depends on the type of cabling and its length. Much higher rates possible such as 115.2 kb/s. The upper limit is determined by maximum cable capacitance of 2500 pF. Standard serial port data speeds are 75, 110, 300, 1200, 2400, 4800, 9600, 19200, 38400, 57600, and 115200 b/s.

CABLE MEDIUM

Any common multiwire cable works. Shielded or unshielded twisted pair works well.

RANGE

Standard states less than 50 feet based on maximum cable capacitance of 2500 pF. Greater lengths are possible with lower capacitance cables. Hundreds of feet are possible as a function of data rate and cable length.

NETWORK CONFIGURATION

- Point-to-point connection between a computer or terminal called the data terminal equipment (DTE) and a peripheral device called the data communications equipment (DCE). One transmitter and one receiver only.
- Unbalanced, single-ended signal plus ground connections.
- Special connectors have been defined, a 25-pin and a 9-pin. See Figure 25.1 showing signals and connections for the popular 9-pin connector. The 25-pin connector is no longer widely used.
- The 9-pin connector is the most common on equipment today.
- Two-wire (TD and GND), three-wire (TD, RD, and GND), and five-wire (TD, RD, RTS, CTS, and GND) connections are the most widely used.

LOGIC LEVELS

Binary 0 (SPACE): +3 to +25V (+5 or +12V typical), Binary 1 (MARK): −3 to −25V (−5 or −12V typical). Figure 25.2.

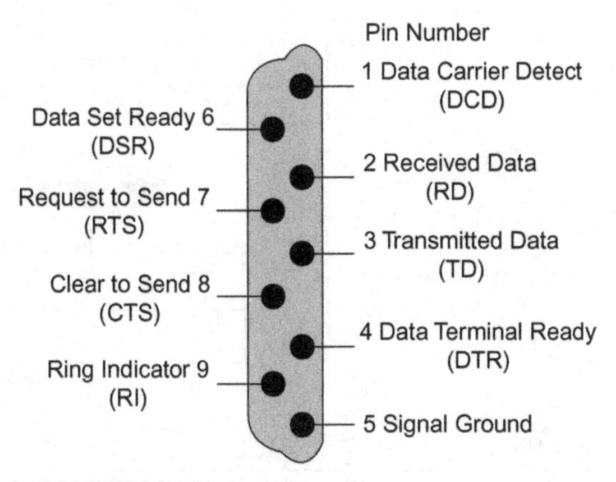

Figure 25.1 9-pin DB9/DE9 RS-232 connector pin out.

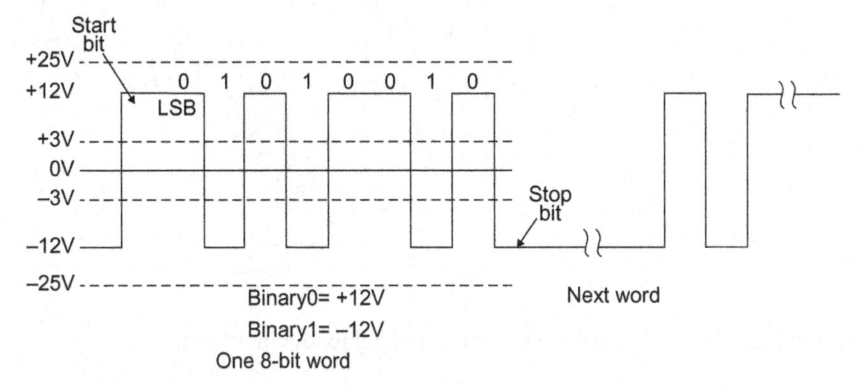

Figure 25.2 Asynchronous data transmission.

PROTOCOL

- Uses standard asynchronous UART transmission with start, stop, and parity bits. Data may be 5 to 8 bits in length. Figure 25.3.
- Original protocol used multiple hard-wired handshaking lines in the cable to carry control signals to synchronize the transmission of bytes between DTE and DCE.
  - Example 1 DTE enables the RTS line to DCE to ask for transmission. DCE enables CTS line to DTE to signal data transfer.
  - Example 3 DCE enables CTS line to tell DTE to transmit. RTS enabled by DTE signals DCE to send data to DTE.
- Software protocols are also used to transmit blocks of data using ASCII control characters. Common protocols include Kermit, XON/XOFF, XMODEM, YMODEM, and ZMODEM.

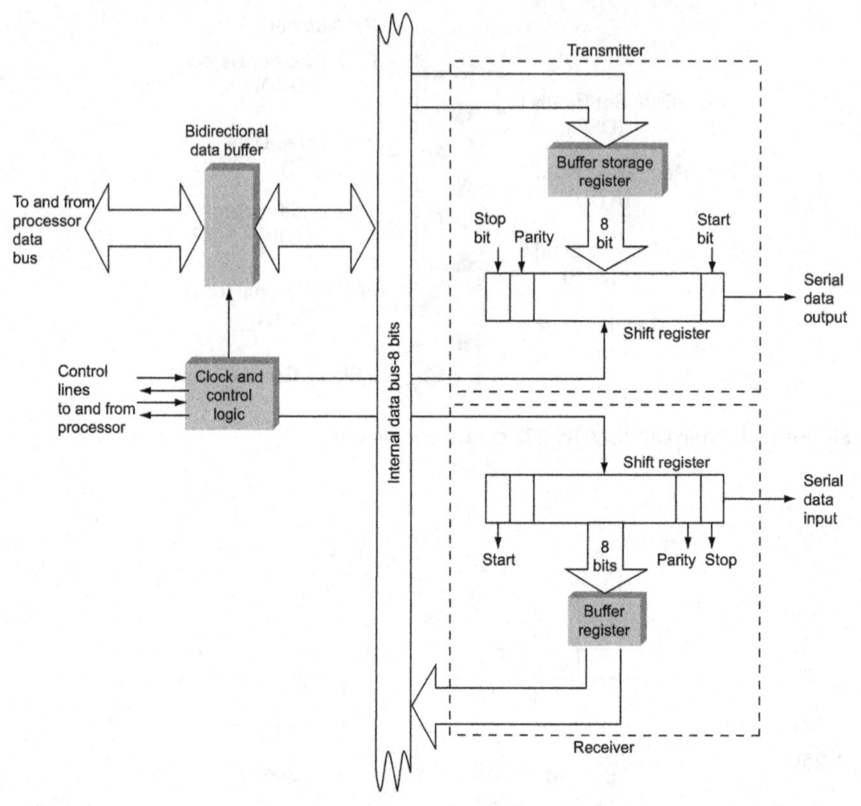

Figure 25.3 The UART is at the heart of most asynchronous interfaces.

- XMODEM Example to transmit a block of data from DTE to DCE:
  - Send ASCII character SOH (start of header).
  - Send block number byte.
  - Send 1's complement of block number byte.
  - Transmit up to 128 bytes of data.
  - Send logical checksum byte signaling end of block.

IC SOURCES

- Analog Devices
- Linear Technology
- Maxim Integrated
- ON Semiconductor
- STMicroelectronics
- Texas Instruments

CHAPTER TWENTY-SIX

RS-422

APPLICATIONS

- Miscellaneous industrial automation uses.
- Broadcast equipment interconnections.
- Extenders/repeaters for RS-232 connections.

SOURCE

Electronics Industry Alliance (EIA).

NATIONAL OR INTERNATIONAL STANDARD

American National Standards Institute (ANSI).

Electronics Industry Alliance/Telecommunications Industry Association, EIA/TIA-422-B.

KEY FEATURES

- Balanced differential data lines greatly minimize noise.
- Higher data rate than RS-232 connections.

DATA RATE

100 kb/s to 10 Mb/s. Depends on cable length. Estimated 100 kb/s at 4000 feet and 10 Mb/s at 40 feet. Guideline: Speed in b/s multiplied by length in meters should not exceed 10^8.

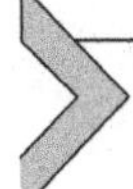

CABLE MEDIUM

Shielded or unshielded twisted pair.

Handbook of Serial Communications Interfaces.
Doi: http://dx.doi.org/10.1016/B978-0-12-800629-0.00026-7

RANGE

Up to 1200 m or 4000 feet.

NETWORK CONFIGURATION

- Point-to-point connection between a computer or terminal called the data terminal equipment (DTE) and a peripheral device called the data communications equipment (DCE). One transmitter and one receiver.
- Multidrop connections with one transmitter and up to ten receivers.
- Simplex operation only where receivers cannot also transmit.
- Balanced signal connections.
- No specific connectors have been designated by the standard. The 9-pin DB9/DE9 connector is sometimes used.

LOGIC LEVELS

Binary 0 (SPACE): +2 to +6V, Binary 1 (MARK): −2 to −6V.

PROTOCOL

- Uses standard asynchronous UART transmission with start, stop, and parity bits. Data may be 5 to 8 bits in length.
- Multiple software protocols are used to transmit blocks of data using ASCII control characters. In addition to proprietary protocols, commonly used protocols include Kermit, XON/XOFF, XMODEM, YMODEM, and ZMODEM. See Chapter 25 on RS-232.

IC SOURCES

- Analog Devices
- Linear Technology
- Maxim Integrated
- ON Semiconductor
- STMicroelectronics
- Texas Instruments

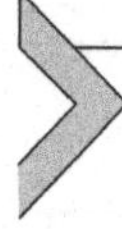

CHAPTER TWENTY-SEVEN

RS-423

APPLICATIONS

Miscellaneous industrial automation uses.

SOURCE

Electronics Industry Alliance (EIA).

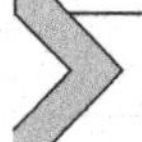

NATIONAL OR INTERNATIONAL STANDARD

American National Standards Institute (ANSI).

Electronics Industry Alliance/Telecommunications Industry Association, EIA/TIA-423-B.

KEY FEATURES

- Unbalanced or single-ended data lines.
- An alternative to RS-232.
- Higher data rate and longer range than RS-232 connections.
- Not widely used.

DATA RATE

100 kb/s maximum. Depends on cable length.

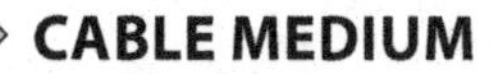

CABLE MEDIUM

Shielded or unshielded twisted pair.

RANGE

Up to 1200 m or 4000 feet.

Handbook of Serial Communications Interfaces.
Doi: http://dx.doi.org/10.1016/B978-0-12-800629-0.00027-9

NETWORK CONFIGURATION

- Point-to-point connection between a computer or terminal called the data terminal equipment (DTE) and a peripheral device called the data communications equipment (DCE). One transmitter and one receiver.
- Multidrop connections with one transmitter and up to ten receivers.
- Simplex where receivers cannot also transmit.
- Unbalanced signal connections.
- No specific connectors have been designated by the standard. The 9-pin DB9/DE9 connector is sometimes used.

LOGIC LEVELS

Binary 0 (SPACE): +2 to +6V, Binary 1 (MARK): −2 to −6V.

PROTOCOL

- Uses standard asynchronous UART transmission with start, stop, and parity bits. Data may be 5 to 8 bits in length.
- Multiple software protocols are used to transmit blocks of data using ASCII control characters. In addition to proprietary protocols, commonly used protocols include Kermit, XON/XOFF, XMODEM, YMODEM, and ZMODEM. See Chapter 25 on RS-232.

IC SOURCES

- Analog Devices
- Linear Technology
- Maxim Integrated
- ON Semiconductor
- STMicroelectronics
- Texas Instruments

CHAPTER TWENTY-EIGHT

RS-485

APPLICATIONS

- Miscellaneous industrial, process control, and commercial networks.
- Building automation.
- Video surveillance.
- Point of sale terminals.

SOURCE

Electronics Industry Alliance (EIA).

NATIONAL OR INTERNATIONAL STANDARD

American National Standards Institute (ANSI), Electronics Industry Alliance/Telecommunications Industry Association, EIA/TIA-485-A.

KEY FEATURES

- Balanced differential data lines greatly minimize noise.
- Duplex transmissions to/from multiple nodes.
- Similar to RS-422 but provides for a greater number of nodes.

DATA RATE

100 kb/s to 10 Mb/s. Depends on cable length. Estimated 100 kb/s at 4000 feet and 10 Mb/s at 40 feet. Guideline: Speed in b/s multiplied by length in meters should not exceed 10^8. Speeds to 35 Mb/s have been achieved. Depends on IC driver specifications.

Handbook of Serial Communications Interfaces.
Doi: http://dx.doi.org/10.1016/B978-0-12-800629-0.00028-0

CABLE MEDIUM

Shielded or unshielded twisted pair.

RANGE

Up to 1200 m or 4000 feet.

NETWORK CONFIGURATION

- Multipoint connections with up to 32 nodes with both transmitters and receivers. See Figure 28.1.
- Half duplex operation.
- Balanced signal connections.
- No specific connectors have been designated by the standard.

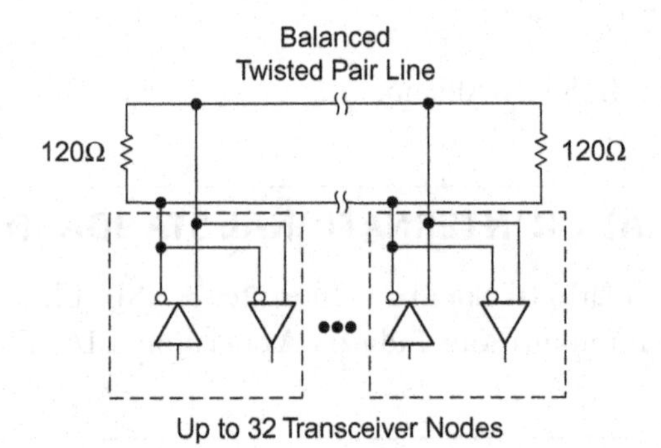

Figure 28.1 RS-485 network configuration.

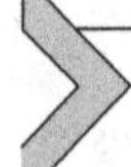

LOGIC LEVELS

Binary 0 (SPACE): +1.5 to +6V, Binary 1 (MARK): −1.5 to −6V.

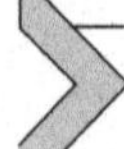

PROTOCOL

- Uses standard asynchronous UART transmission with start, stop, and parity bits. Data may be 5 to 8 bits in length.
- Multiple software protocols are used to transmit blocks of data using ASCII control characters. Multiple proprietary protocols. See Chapter 25 on RS-232.

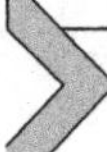

IC SOURCES

- Analog Devices
- Linear Technology
- Maxim Integrated
- ON Semiconductor
- STMicroelectronics
- Texas Instruments

CHAPTER TWENTY-NINE

System Management Bus (SMB)

APPLICATIONS

Used in computers and laptops for control of power monitor and control ICs, temperature monitors, battery chargers, related mechanical devices.

SOURCE

Developed by Intel Corporation in 1995.

NATIONAL OR INTERNATIONAL STANDARD

Maintained by Smart Battery Systems Implementers Forum (smbus@sbs-forum.org) and System Management Interface Forum (SM-IF).

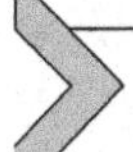

KEY FEATURES

- Derivative of and mostly compliant with I^2C bus specifications. See Chapter 13.
- On–off control on PC motherboards of power devices and mechanical devices like fans.

DATA RATE

10 to 100 kb/s.

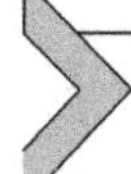

CABLE MEDIUM

Copper connections on a printed circuit board.

Handbook of Serial Communications Interfaces.
Doi: http://dx.doi.org/10.1016/B978-0-12-800629-0.00029-2

RANGE

Less than 1 m.

NETWORK CONFIGURATION

Same as I^2C plus an interrupt line.

LOGIC LEVELS

Binary 0: <0.8V, Binary 1: >2.1V

PROTOCOL

- Same as I^2C with some exceptions including the following:
  - Slave devices must always acknowledge their own addresses.
  - Data transfer formats are a subset of the I^2C specifications.
  - Includes a one-byte CRC-8 checksum at the end of each transmission.
  - Adds a SMBALERT# interrupt line to the bus that slaves can use to notify the master of key events.
  - A 35 ms time-out feature that causes all devices to be reset inferring a problem on the bus.

IC SOURCES

- Cadex
- Intel
- Linear Technology
- Maxim Integrated
- NXP
- Texas Instruments
- Xilinx

CHAPTER THIRTY

T1/E1

APPLICATIONS

Primary digital connection between telephone central offices and some customer premises. Applications include:

- Digital voice telephony
- Internet connectivity
- Wide area networking
- Dedicated data links

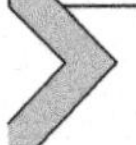

SOURCE

Telephone industry. T1 (United States, Canada), E1 (Europe and others).

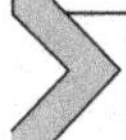

NATIONAL OR INTERNATIONAL STANDARD

- International Telecommunications Union (ITU) G.703 and G.704 for E1.
- American National Standards Institute (ANSI) T1.403 for T1.

KEY FEATURES

- A time division multiplex scheme that encodes multiple telephone calls on a single connection.
- T1 uses 24 8-bit time slots for voice signals sample at an 8 kHz rate.
- E1 uses 32 8-bit time slots for voice signals sampled at an 8 kHz rate.
- Synchronous transmission.

Handbook of Serial Communications Interfaces.
Doi: http://dx.doi.org/10.1016/B978-0-12-800629-0.00030-9

DATA RATE

- T1: 1.544 Mb/s
- E1: 2.048 Mb/s

CABLE MEDIUM

Two shielded twisted pairs (T1 and E1). Two 75-ohm coax cables (E1). Connectors are RJ-48 (similar to RJ-45 Ethernet connectors) or DB9/DE9 RS-232 connectors. Two BNC coax connectors for E1.

RANGE

Several thousand meters. Up to a maximum of about 12,000 feet.

NETWORK CONFIGURATION

- Point-to-point connections.
- Full duplex with two data paths.
- Differential connections.
- Clock derived from data signal.

LOGIC LEVELS

Unique line coding used for long-range reliability, ease of clock recovery, and elimination of DC build up on the line:

- Alternate mark inversion (AMI) – Binary 1 is alternately +3V then −3V, etc. (T1)
- Bipolar 8-Zero Substitution (B8ZS) – Similar to AMI but with an added binary 1 after eight sequential binary 0s to ensure reliable clock recovery. (T1)
- High Density Bipolar 3 (HDB3) - Similar to B8ZS but with an added binary 1 after four sequential binary 0s to ensure reliable clock recovery. (E1)

  Binary 1 (MARK): +3V or −3V, Binary 0 (SPACE): 0V

PROTOCOL

Figure 30.1 shows a typical T1 data frame with 24 8-bit time slots to be filled with voice signals sampled at 8 kHz or data. A single extra bit is added for synchronization. This produces a 193-bit frame.

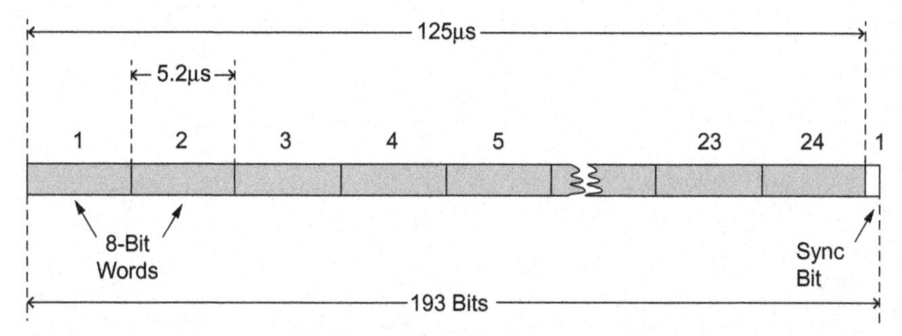

Figure 30.1 T1 frame format that time multiplexes 24 8-bit words into a serial data block at 1.544 Mb/s.

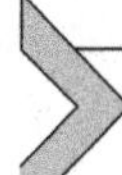

IC SOURCES

- FLEX
- Maxim Integrated
- NEC
- Shenzhen Jin Pengyue Electronics

SECTION II

Medium-Speed Interfaces (10 Mb/s to 1 Gb/s)

CHAPTER THIRTY-ONE

Ethernet

APPLICATIONS

- Local Area Networks
- Metropolitan Area Networks
- Wide Area Networks
- Industrial Networks
- Backplanes

SOURCE

Originally developed in 1973 at Xerox Palo Alto Research laboratories.

NATIONAL OR INTERNATIONAL STANDARD

Institute of Electrical and Electronic Engineers (IEEE) 802.3

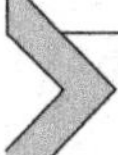

KEY FEATURES

- Most widely adopted wired networking technology.
- Most PCs and laptops feature an Ethernet port.
- Multiple speed levels (10 Mb/s to 100 Gb/s)
- Twisted pair and fiber physical medium options.
- Wireless versions available (IEEE 802.11)

DATA RATE

Available rates: 10 and 100 Mb/s. 1, 2.5, 5, 10, 25, 50, 40, and 100 Gb/s.

NOTE: The 10 Mb/s, 100 Mbps, and 1 Gb/s versions are covered here. The 10 to 100 Gb/s versions are covered in the High-Speed Interface section.

Handbook of Serial Communications Interfaces.
Doi: http://dx.doi.org/10.1016/B978-0-12-800629-0.00031-0

CABLE MEDIUM

- Unshielded twisted pair (UTP): CAT5/5e/6/7.
- RJ-45 connector.
- Original coax cable versions (RG-8/U and RG-58/U) no longer used.
- 9 μm single mode fiber (SMF), 50 and 62.5 μm multimode fiber (MMF) such as OM3 and OM4.

RANGE

Typical range 100 m without repeaters or other regeneration.

NETWORK CONFIGURATION

- Full duplex over two UTP pairs (10 and 100 Mb/s).
- Four UTP pairs used in 1 and 10 Gb/s versions.
- Differential signaling.
- Logical bus topology, physical star.
- Carrier-sense multiple access with collision detection (CSMA/CD) access method.
- Manchester encoding for 10 Mb/s. MLT-3 for 100 Mb/s, PAM5 for 1 Gb/s.
- Fiber optic versions use 8B/10B encoding.

LOGIC LEVELS

Binary 0: 0V, Binary 1: −2V for 10 Mb/s version. ±1V for MLT-3.

PROTOCOL

- Layers 1 and 2 of the OSI model.
- See protocol frame in Figure 31.1.

Number of bytes per field

7	1	6	6	2	46 to 1500	4
Preamble	Start Frame Delimiter	Destination Address	Source Address	Length	Data	Frame Check Sequence

Figure 31.1 Ethernet (802.3) protocol frame.

IC SOURCES

- Broadcom
- Intel
- Marvell
- SMSC
- Vitesse Semiconductor

CHAPTER THIRTY-TWO

FireWire

APPLICATIONS

- Connecting audio and video equipment (cameras, DVRs, monitors, editors/mixers, etc.)
- Primary application is data download from camcorders.
- Connections to remote disk drive data storage devices.
- Some satellite receivers and MIDI synthesizers.

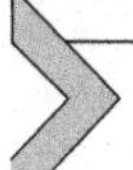

SOURCE

Apple developed FireWire in the 1980s and deployed it in Macintosh computers in the 1990s.

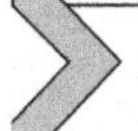

NATIONAL OR INTERNATIONAL STANDARD

Institute of Electrical and Electronic Engineers (IEEE) 1394–2008.

KEY FEATURES

- High speed serial interface primarily for video data transfer.
- A Sony version is called i.LINK.
- A Texas Instrument version is called Lynx.
- Hot pluggable and plug-and-play capable.
- Once used to replace parallel Small Computer System Interface (SCSI) interconnections.
- Similar to USB.
- Optical and backplane versions defined but not commonly available.
- No longer widely used in Apple computers.

Handbook of Serial Communications Interfaces.
Doi: http://dx.doi.org/10.1016/B978-0-12-800629-0.00032-2

DATA RATE

100 (98.304), 200 (196.608), 400 (393.216), 800 (786.432) Mb/s, 1.6 and 3.2 Gb/s.

CABLE MEDIUM

Six wire shielded twisted pair cable with two differential pairs plus DC power and ground. The i.LINK version uses only the two differential pairs without the power wires. A nine wire cable is used on the 1.6 and 3.2 Gb/s versions.

CONNECTORS

Special connectors defined. See Figure 32.1.

RANGE

Maximum cable length is 4.5 m or 14.8 feet. Up to 16 devices may be chained for a total length of 72 m.

NETWORK CONFIGURATION

- Peer-to-peer connection is the most common.
- Up to 63 devices may be daisy-chained.
- Bidirectional data transfer.
- Asynchronous and isosynchronous data transfer options.

LOGIC LEVELS

Binary 1: >2V, Binary 0: <0.8V. + 3.3V supply typical, LPTTL/HCMOS.

PROTOCOL

- Multiple protocols available depending upon application (automotive, industrial, networking, camcorders, etc.)
- Three-layer OSI format: transaction (layer 3), data link (layer 2) and physical (layer 1).
- Packet-based frame.

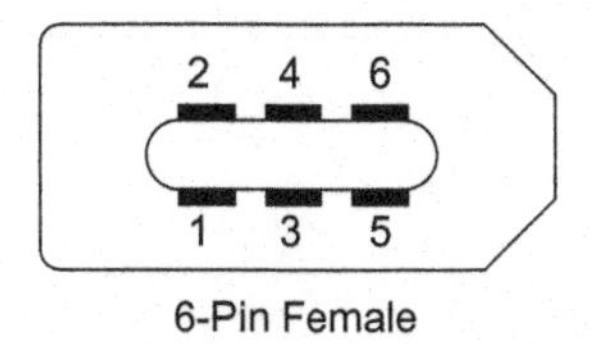

(a) Standard FireWire Connector

Pin	Name
1	+DC Power (up to 30V)
2	Ground
3	-TPB
4	+TPB
5	-TPA
6	+TPA

TPA = Twisted Pair A
TPB = Twisted Pair B

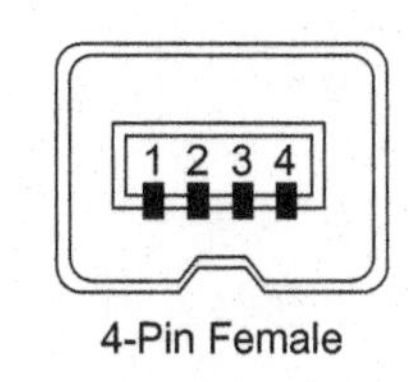

(b) I-LINK Connector

Pin	Name
1	-TPB
2	+TPB
3	-TPA
4	+TPA

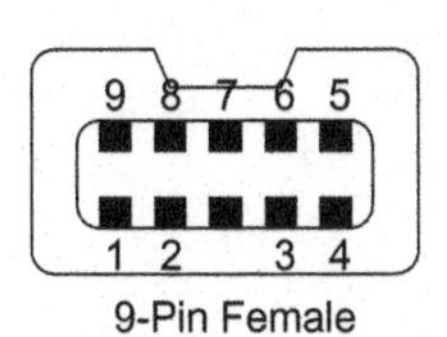

(c) Newer Connector for High Speeds

Pin	Name
1	-TPB
2	+TPB
3	-TPA
4	+TPA
5	Pair A Ground
6	Power Ground
7	No Connection
8	+DC Power
9	Pair B Ground

Figure 32.1 FireWire Connector pin outs.

- Isochronous data transfer option that ensures real-time data transfer for critical video data without buffering delays. Special data strobe encoding ensures synchronization.
- An arbitration mode decides which node has the priority to transmit.
- Operating system support from Apple Macintosh iOS, FreeBSD, Linux, Microsoft Windows.

IC SOURCES

- Adaptec
- Cirrus Logic
- Maxim Integrated
- LSI
- Symbios
- Texas Instruments

CHAPTER THIRTY-THREE

Joint Test Action Group (JTAG)

APPLICATIONS

Testing of printed circuit boards and large-scale integrated circuits in a production environment.

SOURCE

Institute of Electrical and Electronic Engineers (IEEE)

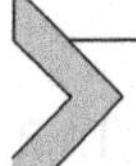

NATIONAL OR INTERNATIONAL STANDARD

IEEE 1149.1, 1149.7, and 1687

KEY FEATURES

- Ability to automate the testing of complex printed circuit boards or large ICs like microprocessors, ASICs, FPGAs, etc.
- Use of boundary scan techniques to test IC pin connections.
- Ability to build internal "instruments" that allow testing of IP sections of complex mixed signal ICs (IJTAG).
- Interface is called the serial test access port (TAP) that allows test programs to be entered and test data extracted.

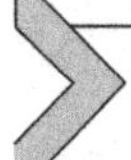

DATA RATE

Usually in the 10 to 50 Mb/s range.

CABLE MEDIUM

Not specified.

Handbook of Serial Communications Interfaces.
Doi: http://dx.doi.org/10.1016/B978-0-12-800629-0.00033-4

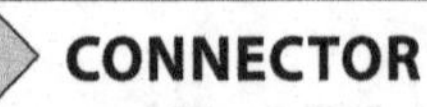

CONNECTOR

Not specified. Usually PCB headers.

RANGE

Not specified.

NETWORK CONFIGURATION

- Point-to-point.
- Daisy chained.
- Test access port (TAP). See Figure 33.1
    - TCK – Test Clock. Data entered on rising edge, data read on trailing edge.
    - TDI – Test Data Input. Serial data test program.
    - TDO – Test Data Output. Serial test results.
    - TMS – Test Mode Select. Determines state of the TAP controller.
    - TRST – Asynchronous reset of system logic.

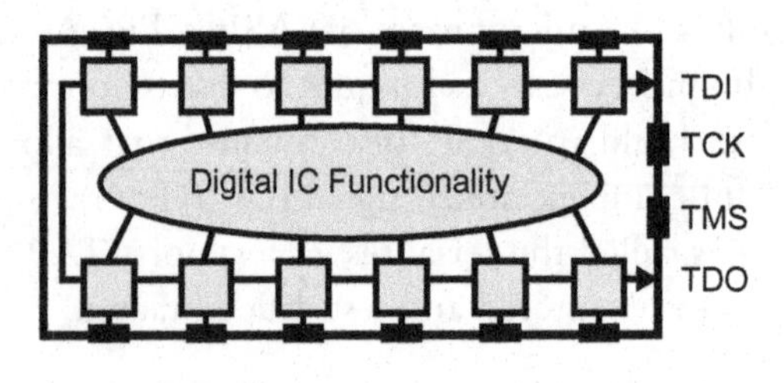

Figure 33.1 A flip flop at each IC test pin can be set or reset. The flip flops are connected to form a shift register that can shift test data in or out.

LOGIC LEVELS

As required by the circuits under test.

PROTOCOL

Determined by the Boundary Scan Description Language (BSDL) used in programming the test operations. Other programming languages used are the Instrument Connectivity Language (ICL) and the Procedural Description Language (PDL) and the Tool Command Language (TCL).

IC SOURCES

Most semiconductor manufacturers of large-scale ICs like processors, ASICs, or FPGAs incorporate JTAG and/or IJTAG.

CHAPTER THIRTY-FOUR

Media Oriented Systems Transport (MOST)

APPLICATIONS

Transport of audio, video, voice, and data signals in automotive vehicles.

SOURCE

Standard Microsystems Corporation (SMSC) now part of Microchip Technology.

NATIONAL OR INTERNATIONAL STANDARD

MOST Cooperation

KEY FEATURES

- A multimedia network optimized for infotainment systems in vehicles.
- Handles an audio/visual media: stereo audio, video, GPS/navigation, data, and control.

DATA RATE

25 (24.8), 50 or 150 Mbps. Bi-phase encoding.

CABLE MEDIUM

Plastic optical fiber (POF) or unshielded twisted pair (UTP)

RANGE

Within a vehicle.

Handbook of Serial Communications Interfaces.
Doi: http://dx.doi.org/10.1016/B978-0-12-800629-0.00034-6

NETWORK CONFIGURATION

- Ring network topology.
- Supports up to 64 devices.
- Synchronous communications supports media streaming without buffering.
- Support for asynchronous transfers.

LOGIC LEVELS

IR light levels or Ethernet wired levels.

PROTOCOL

- Timing master controls all operations to timing followers.
- Complete OSI model for all seven layers.
- Data transferred in blocks of 16 frames.
- Each frame has 512 bits of which 60 bytes is data.
- Frame rate is 44.1 kbps.

IC SOURCES

- Analog Devices
- Microchip Technology

CHAPTER THIRTY-FIVE

Serial Peripheral Interface (SPI)

APPLICATIONS

- Chip-to-chip connections on a printed circuit board (PCB) or between PCBs.
- Microcontroller to peripheral devices like memory chips, ADC, DAC, displays, sensors, etc.

SOURCE

Originated by Motorola now Freescale.

NATIONAL OR INTERNATIONAL STANDARD

None

KEY FEATURES

- High-speed full duplex bus.
- Three-wire bus with input, output, and clock lines plus slave select lines.
- Synchronous data transfer with single clock.
- Superset of an older version known as Microwire. See Chapter 18.

DATA RATE

Unspecified. Typically 20 Mb/s up to about 100 Mb/s.

CABLE MEDIUM

Copper pattern on PCB or short ribbon cable between PCBs.

Handbook of Serial Communications Interfaces.
Doi: http://dx.doi.org/10.1016/B978-0-12-800629-0.00035-8

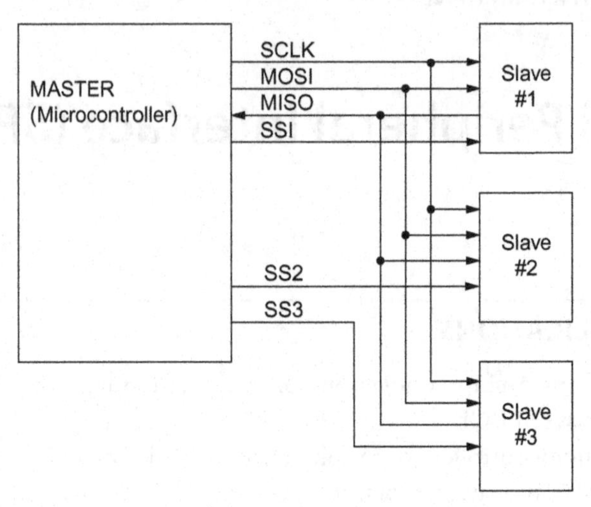

Figure 35.1 Serial Peripheral Interface (SPI).

RANGE

Up to 1 m but typically less than one foot.

NETWORK CONFIGURATION

- Master-slave connections where the master is usually an embedded controller and slaves are peripheral chips. See Figure 35.1.
- Three lines: Master out slave in (MOSI), master in slave out (MISO), and clock (SCLK).
- Slaves must have a slave select (SSn) line usually implemented with a GPIO bus line from the master microcontroller.
- Slaves may be daisy-chained and a single common slave select line is used.

LOGIC LEVELS

TTL or CMOS compatible: Binary 0: <0.4V, Binary 1: >2.4V.

PROTOCOL

- Protocol usually proprietary to the specific application or operation of the slave ICs.

- Word size is typically 8-bits but 12, 16, and other bit word sizes can be accommodated.
- Data transmission sequence:
  - Master transmits a binary 0 to the selected slave.
  - Slave sends data to master or master sends data to slave. MSB first.
  - Different modes of operation specify whether data is transmitted or captured by the leading or trailing edge of the clock.

IC SOURCES

- Analog Devices
- ARM
- Atmel
- Freescale
- Intel
- Maxim Integrated
- Microchip Technology
- Texas Instruments

CHAPTER THIRTY-SIX

Universal Serial Bus (USB)

APPLICATIONS

- All purpose serial interface for personal computer peripherals such as disk drives, printers, mice, keyboards, joy sticks, etc.
- Consumer electronics such as video cameras, digital cameras, video games, etc.
- Flash storage devices (USB drives).
- Battery chargers for cell phones and tablets.

SOURCE

Original source: Companies Compaq, DEC, IBM, Intel, Microsoft, NEC, and Nortel.

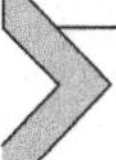

NATIONAL OR INTERNATIONAL STANDARD

Universal Serial Bus Implementer's Forum (USB-IF).

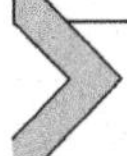

KEY FEATURES

- Most widely used interface other than RS-232.
- Standardized bus, connectors, and transmission specifications and protocol.
- Widely used on virtually all PCs and peripherals and consumer equipment
- Plug and play, hot pluggable.
- Includes DC power connections (+5V). Power level depends on version: 0.75 to 100 watts.
- Wide speed ranges from 1.5 Mbps to 10 Gbps.

Handbook of Serial Communications Interfaces.
Doi: http://dx.doi.org/10.1016/B978-0-12-800629-0.00036-X

DATA RATE

Maximum speeds:

- Version 1.1 1.5 or 12 Mbps
- Version 2.0 480 Mbps
- Version 3.1 5 or 10 Gbps

CABLE MEDIUM

Twisted pair with shield. Four wires with shield for Versions 1.0–2.0 and ten wires with three data pairs, ID pin wire, DC power, and two grounds.

CONNECTORS

Multiple types defined and implemented. The most common are the Type A and B shown in Figure 36.1 with pin out information. Figure 36.2 shows the Micro-B connector for version 3.0.

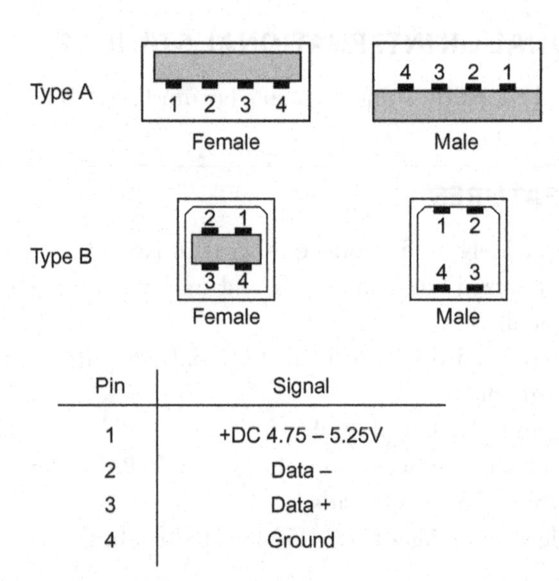

Pin	Signal
1	+DC 4.75 – 5.25V
2	Data –
3	Data +
4	Ground

Figure 36.1 Most common USB connectors.

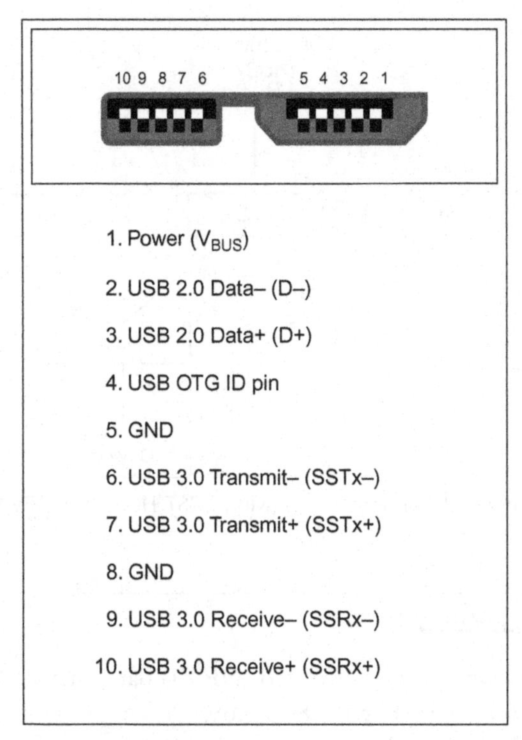

Figure 36.2 Micro-B USB connector.

RANGE

Maximum cable length: 5 m for most versions; 3 m for version 3.0. Repeaters and extender cables are available to extend range to 100 m.

NETWORK CONFIGURATION

- Serial multi-level star topology. See Figure 36.3.
- One master host controller and multiple slave nodes. Host node is a hub that is a bus. A hub supports multiple nodes. Hubs can be chained.
- Maximum number of nodes 127.
- Differential pair data paths. One pair for versions 1.0–2.0, three pairs for version 3.0/3/1.

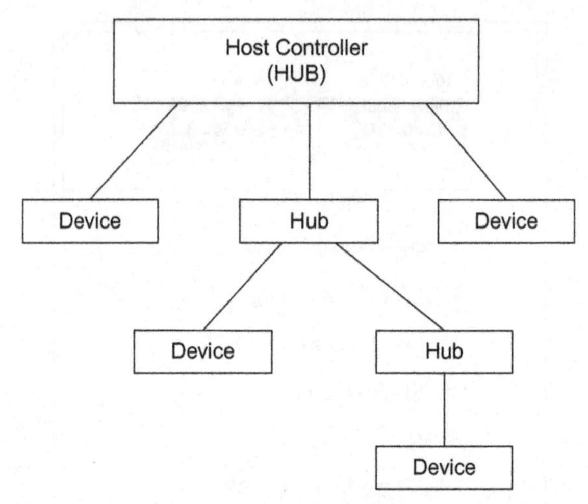

Figure 36.3 USB network connections showing HOST, HUBS, and DEVICE nodes.

LOGIC LEVELS

- Non-return to zero inverted (NRZI) data format.
- Low: 0 to 0.3V, High: 2.8 to 3.6V.
- Alternately for faster data versions: Low: 0 to 10 mV, High: 360 to 440 mV.

PROTOCOL

- Half duplex for most versions.
- Full duplex for version 3.0/3.1.
- Host initiates all communications.
- Transmission is by packets. Multiple packets make up a frame.
- Multiple packet types designate operations.
  - Token packets define operations.
  - Data packets transfer data, up to 1024 bytes.
  - Status packets are handshake packets to confirm the operation.
  - Special packets define auxiliary operations.
- All packets start with an 8-bit sync field, a packet ID (PID) field to define the type of packet, and end with a 3-bit end of packet (EOP) field.
- Token packets have a 7-bit address, token packets have a 5-bit CRC, data packets have a 16-bit CRC.

IC SOURCES

- Atmel
- Exar
- Fairchild
- Freescale
- FTDI
- Fujitsu
- Intel
- NXP
- ON Semiconductor
- SMSC
- ST Ericsson
- STMicroelectronics
- Texas Instruments

SECTION III

High-Speed Interfaces (1–100 Gb/s)

CHAPTER THIRTY-SEVEN

100 Gigabit Ethernet Attachment Unit Interface (CAUI)

APPLICATIONS

A chip-to-chip or chip-to-module interface between Ethernet Media Access Controller (MAC) layer 2 circuitry and the physical layer 1 module used by the main Ethernet medium (fiber optic cable).

SOURCE

Institute of Electrical and Electronic Engineers (IEEE)

NATIONAL OR INTERNATIONAL STANDARD

IEEE 802.3ba and extensions.

KEY FEATURES

- An interface that reduces the number of data connections between chips and modules in an Ethernet system.
- Reduces the 100 Gigabit Media Independent Interface (CGMII) that features two 32-bit data buses plus two 4-bit control buses to ten 10 Gigabit serial interfaces.
- Another version (CAUI-4) reduces the 100 Gigabit Media Independent Interface (CGMII) that features two 32-bit data buses plus two 4-bit control buses to four 25 Gigabit serial interfaces.

DATA RATE

100 Gb/s

Handbook of Serial Communications Interfaces.
Doi: http://dx.doi.org/10.1016/B978-0-12-800629-0.00037-1

CABLE MEDIUM

PCB traces. Back planes

RANGE

Up to 50 cm.

NETWORK CONFIGURATION

- CAUI – ten 10.3125 Gb/s lanes
- CAUI-4 – four 25 Gb/s lanes
- 64b/66b encoding.

LOGIC LEVELS

Binary 0: −0.3V, Binary 1: +0.3V.

PROTOCOL

- A physical layer standard only.
- Uses the standard Ethernet protocol and 100 Gigabit Media Independent Interface (CGMII) at the MAC layer.

IC SOURCES

- Altera
- Xilinx

CHAPTER THIRTY-EIGHT

Common Electrical Interface – 28 Gigabit (CEI-28G)

APPLICATIONS

Chip-to-chip, board-to-board, or chip-to-module interface to achieve 100 Gb/s data rate in Ethernet, Fibre Channel, Infiniband, or Optical Transport Network (OTN) systems.

SOURCE

Optical Internetworking Forum (OIF)

NATIONAL OR INTERNATIONAL STANDARD

Optical Internetworking Forum (OIF) CEI-28G-VSR (very short reach).

KEY FEATURES

- Four serial lanes of 28 Gb/s to achieve 100 Gb/s+.
- Configuration to achieve BER of 10^{-15}.
- OIF is developing a similar standard for four lanes of 56 Gb/s to achieve a gross data rate of 400 Gb/s for future networks. (OIF CEI-56G-VSR)

DATA RATE

100 Gb/s

CABLE MEDIUM

PCB traces. Back planes.

Handbook of Serial Communications Interfaces.
Doi: http://dx.doi.org/10.1016/B978-0-12-800629-0.00038-3

RANGE

Inches.

NETWORK CONFIGURATION

Point-to-point.

LOGIC LEVELS

1200 mV peak-to-peak, differential.

PROTOCOL

None specified. Physical layer interface only.

IC SOURCES

- Altera
- Gennum
- IBM
- Inphi
- Xilinx

CHAPTER THIRTY-NINE

Common Public Radio Interface (CPRI)

APPLICATIONS

Connection of remote radio head (RRH) to the radio equipment control (REC) unit in a cellular telephone base station. Also used in some distributed antenna systems (DAS).

SOURCE

Industry Initiative Common Public Radio Interface.

NATIONAL OR INTERNATIONAL STANDARD

Joint industry consortium made up of companies Alcatel-Lucent, Ericsson, Huawei, NEC, Nokia Siemens Networks, and Nortel.

KEY FEATURES

Allows mounting the radio equipment of a cellular base station on the tower near the antenna to eliminate the high cost and attenuation of the usual long coaxial transmission lines.

DATA RATE

- Multiple rates can be selected including 614.4 Mbps, 1.2288, 2.458, 3.072, 4.915, 6.144, and 9.830 Gbps.
- 10^{-12} BER required without FEC.

Handbook of Serial Communications Interfaces.
Doi: http://dx.doi.org/10.1016/B978-0-12-800629-0.00039-5

CABLE MEDIUM

- Fiber optic cable. Options for both single mode fiber (SMF) and multimode fiber (MMF) are available.
- Physical layer not specified but can use Gigabit Ethernet, 10 Gigabit Ethernet, Fibre Channel, or InfiniBand. Transceivers can use the standard SFP or SFP+ form factor.
- Twisted pair copper cables like those used in Gigabit Ethernet are an option but not common.

RANGE

Form several hundred feet to 80 km.

NETWORK CONFIGURATION

- Direct connection between the radio equipment control (REC) unit in a base station and the remote radio head (RRH) is the most common.
- Multiple RRH may be served by one REC base station using a daisy-chained, ring, tree, branch, or star network configuration.

LOGIC LEVELS

IR light levels.

PROTOCOL

- Time-division multiplexing is used in transmitting control and data words.
- A basic CPRI frame consists of 16 words that may vary in length from 8 to 128 bits each.
- Higher data rates use longer words.
- 8B/10B coding is used.
- The first of the 16 words is a control word.
- The remaining 15 words in a frame are the digital I and Q signals to and from the baseband circuits in the REC to the RRH.
- Multiple basic frames are combined into hyperframes that begin with a sync word.

IC SOURCES

- Altera
- LSI
- Xilinx

CHAPTER FORTY

DisplayPort (DP)

APPLICATIONS

Connect computers or other devices to a video monitor or projector.

SOURCE

Video Electronics Standards Association (VESA)

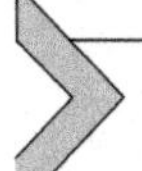

NATIONAL OR INTERNATIONAL STANDARD

Video Electronics Standards Association (VESA)

KEY FEATURES

- Replaces older and slower video interfaces like DVI, FPD-Link, and VGA.
- Competes with HDMI in some applications.
- Supports the transport of very high definition color digital video.
- Handles multiple color formats using 6, 8, 10, 12, and 16-bit color words.
- Supports the transmission of digital audio.
- DC voltage on cable (Usually 3.3V but up to 16V and 500mA.)
- Hot pluggable.
- Dual mode (DisplayPort++) version

DATA RATE

1.296, 2.16, or 4.32 Gb/s per channel. Maximum of 17.28 Gb/s with 4 channels.

Handbook of Serial Communications Interfaces.
Doi: http://dx.doi.org/10.1016/B978-0-12-800629-0.00040-1

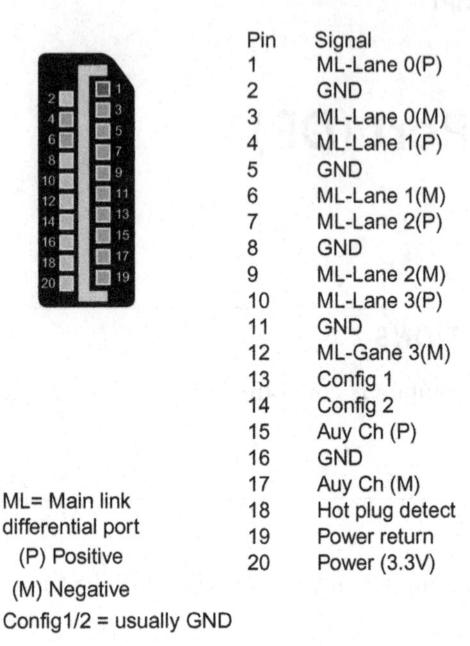

Pin	Signal
1	ML-Lane 0(P)
2	GND
3	ML-Lane 0(M)
4	ML-Lane 1(P)
5	GND
6	ML-Lane 1(M)
7	ML-Lane 2(P)
8	GND
9	ML-Lane 2(M)
10	ML-Lane 3(P)
11	GND
12	ML-Gane 3(M)
13	Config 1
14	Config 2
15	Auy Ch (P)
16	GND
17	Auy Ch (M)
18	Hot plug detect
19	Power return
20	Power (3.3V)

Figure 40.1 DisplayPort connector pinout.

CABLE MEDIUM

20-wire cable, with four differential pairs for video and audio.

CONNECTOR

Unique 20-pin connector. See Figure 40.1.

RANGE

Maximum passive cable length for maximum speed is 2 m. Maximum range with an active cable is 33 m.

NETWORK CONFIGURATION

- Point-to-point connections between video/audio source and the destination device.
- Support for daisy chaining.
- Four differential lanes, simplex.

- 1 to 8 time multiplexed digital audio channels with 16 or 24-bit audio using linear PCM and sampling rates from 32 to192 kb/s.
- Auxiliary channel for management and control, half duplex.

LOGIC LEVELS

NRZ, AC coupled, 400 mV peak-to-peak.

PROTOCOL

- Uses packets to transport data with embedded clock.
- 8B/10B encoding.

IC SOURCES

- Analog Devices
- IDT
- Intel
- Intersil
- NXP
- Parade Technologies
- Texas Instruments

CHAPTER FORTY-ONE

Gigabit Ethernet (GE)

APPLICATIONS

- Local Area Networks
- Metropolitan Area Networks
- Wide Area Networks
- Industrial Networks
- Data center connectivity
- Backplanes

SOURCE

Originally developed in 1973 at Xerox Palo Alto Research laboratories.

NATIONAL OR INTERNATIONAL STANDARD

Institute of Electrical and Electronic Engineers (IEEE) 802.3ae and 802.3ba.

KEY FEATURES

- Most widely adopted wired networking technology.
- Most PCs and laptops feature an Ethernet port.
- Multiple speed levels (10 to 100 Gb/s).
- Twisted pair and fiber optical physical medium options.
- 25 Gb/s twisted pair and 400 Gb/s fiber versions are under development.

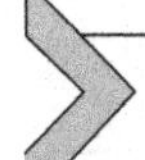

DATA RATE

Available rates: 10, 40, and 100 Gb/s.

Handbook of Serial Communications Interfaces.
Doi: http://dx.doi.org/10.1016/B978-0-12-800629-0.00041-3

Note: The 10 Mb/s, 100 Mbps, and 1 Gb/s versions were covered in Chapter 32. The 10 to 100 Gb/s versions are covered in this chapter.

CABLE MEDIUM

- Unshielded twisted pair (UTP): CAT5e/6/7.
- RJ-45 connector.
- Twin-ax coaxial cable.
- 9 μm single mode fiber (SMF), 50 and 62.5 μm multimode fiber (MMF) such as OM3 and OM4.

RANGE

Multiple ranges from several feet to many kilometers depending on the version (Tables 41.1 and 41.2).

Table 41.1 40G and 100G versions of 802.3ba showing medium and range

Physical Layer (medium)	Range (m) Up to...	40GE	100GE
Backplane	1 m	40GBASE-KR4	100GBASE-KP4
Improved backplane	1 m		100GBASE-KR4
Twin-ax copper coax cable	7 m	40GBASE-CR4	100GBASE-CR10
CAT8 twisted pair	30 m	40GBASE-T	
MMF (OM3)	100 m	40GBASE-SR4	100GBASE-SR10
MMF (OM4)	125 m	40GBASE-SR4	100GBASE-SR10
SMF	2 km	40GBASE-FR	
SMF	10 km	40GBASE-LR4	100GBASE-LR4
SMF	40 km		100GBASE-ER4

Table 41.2 100GE versions with data rates and ranges

Version	Medium	Lanes	Data Rate (Gb/s)	Range
100GBASE-CR10	Twin-ax cable	10	10.3125	<1 m
100GBASE-CR4	Twin-ax cable	4	25.78125	<7 m
100GBASE-SR10	MMF	10	10.3125	100 m
100GBASE-SR4	MMF	4	25.78125	100 m
100GBASE-LR4	SMF WDM	4 λ	25.78125	10 km
100GBASE-ER4	SMF WDM	4 λ	25.78125	30–40 km
100GBASE-KR4	Backplane	4	25.78125	<1 m
100GNASE-KP4	Backplane	4	25.78125	<1 m

λ = IR light wavelength.

NETWORK CONFIGURATION

- Multiple physical layer versions.
- Differential signaling.
- Logical bus topology, physical star.
- Full duplex operation.
- Reed-Solomon forward error correction.
- BER 10^{-12}.
- Fiber optic versions use 64B/66B encoding.

LOGIC LEVELS

IR light levels at 850, 1200–1300, or 1550 nm, NRZ.

PROTOCOL

- Layers 1 and 2 of the OSI model.
- Same MAC layer but different PHY layers.
- See protocol frame in Figure 41.1.

IC SOURCES

- Broadcom
- Intel
- Marvell
- SMSC
- Vitesse Semiconductor

Number of Bytes per Field

7	1	6	6	2	46 to 1500	4
Preamble	Start Frame Delimiter	Destination Address	Source Address	Length	Data	Frame Check Sequence

Figure 41.1 Gigabit Ethernet frame format.

CHAPTER FORTY-TWO

Fibre Channel (FC)

APPLICATIONS

- Primary application is storage area networks (SANs).
- Connections of peripheral devices to mainframes and supercomputers.

SOURCE

American National Standards Institute (ANSI), T11 Technical Committee of the International Committee for Information Technology Standards (INCITS).

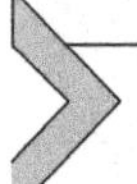

NATIONAL OR INTERNATIONAL STANDARD

Fibre Channel Industry Association (FCIA)

KEY FEATURES

- An optimized way to connect servers and large computers to massive storage units.
- Highly secure since the SAN is isolated from the Internet connections.
- Competes with Ethernet-based Internet Small Computer Systems Interface (iSCSI) SAN alternative.
- High data rates to 16 Gb/s.
- 8 B/10 B or 64 B/66 B encoding.
- Fibre Channel over Internet Protocol (FCIP) encapsulates FC frames into packets for use with TCP/IP for Internet connectivity.

Handbook of Serial Communications Interfaces.
Doi: http://dx.doi.org/10.1016/B978-0-12-800629-0.00042-5

DATA RATE

- Multiple rates from 1.0625 Gb/s (1 Gigabit Fibre Channel-1GFC) to 14.025 Gb/s (16GFC).
- Intermediate rates of 2.125, 4.25, 8.5, and 10.52 Gb/s.
- 10GFC and 16GFC versions use 64 B/66 B encoding.
- Higher rates to 28.05 Gb/s and 112.2 Gb/s are proposed and being developed.

CABLE MEDIUM

- Primarily fiber optic cable both 50 and 62.5 μm multimode fiber (MMF) and 9 μm single mode fiber (SMF).
- Copper options include coax or shielded twisted pair with DB9 connector.

RANGE

- Several hundred meters with MMF or copper.
- 10 to 40 km with SMF.

NETWORK CONFIGURATION

- Host computers are known as initiators while the storage units are called targets.
- Servers and storage units are connected to the SAN by way of an interface called a host bus adapter (HBA).
- Three topology options: See Figure 42.1.
  - Direct point-to-point.
  - Arbitrated loop (Up to 127 nodes).
  - Switch fabric. (Maximum nodes -2^{24}.) The switch fabric is the most common.

LOGIC LEVELS

IR light levels (850, 1310, 1490, 1550 nm options depend upon range).

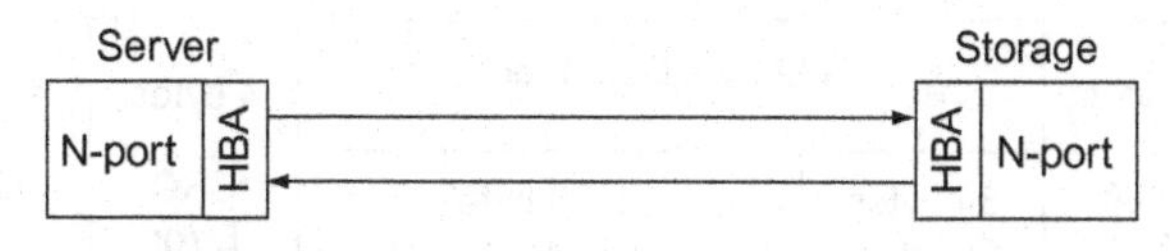

(a) Point-to-Point

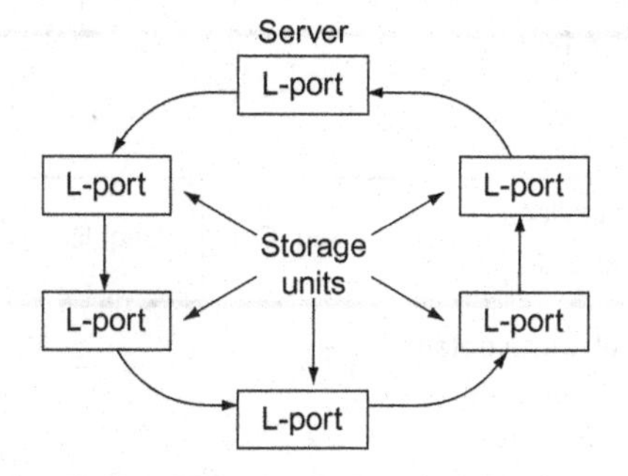

(b) Arbitrated loop or token ring.

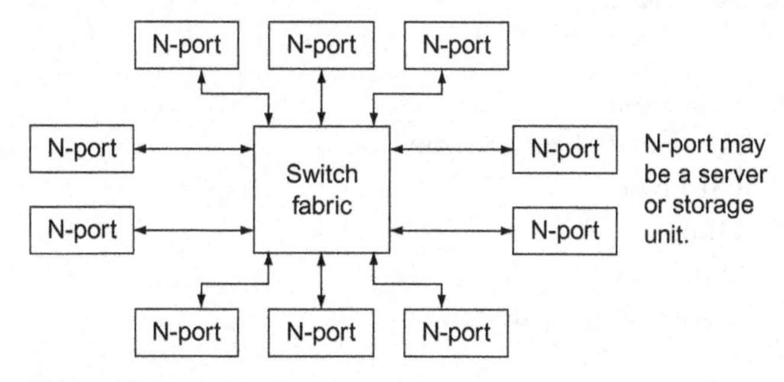

(c) Switch fabric.

Figure 42.1 Fibre Channel topology options. Switch fabric is the most common.

PROTOCOL

- 2148-byte packet or frame. 2112 data bytes, 32-bit CRC. See Figure 42.2.
- Layered protocol similar to but not following the OSI model.
  - FC-0 – Physical layer.
  - FC-1 – Data link layer, line coding.
  - FC-2 – Network layer, frame formatting, flow control.
  - FC-3 – Common services layer, multiple ports, other future options.
  - FC-4 – Protocol-mapping layer, mapping to SCSI, IP, ATM, etc.

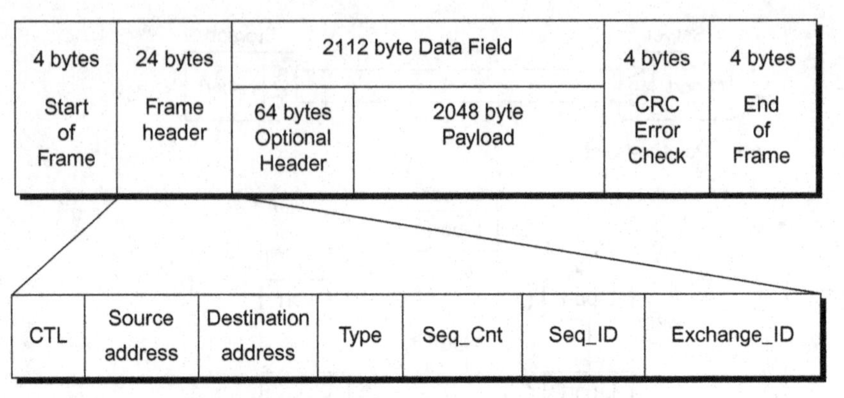

Figure 42.2 Fiber Channel frame details.

IC SOURCES

- Adaptec
- Broadcom
- Hewlett Packard Tachyon
- LSI Logic
- QLogic
- Symbious
- Vitesse Semiconductor

CHAPTER FORTY-THREE

High-Definition Multimedia Interface (HDMI)

APPLICATIONS

Connecting digital video and audio devices together including HDTV sets, DVRs, Blu-Ray players, set top boxes, digital video games, camcorders, digital cameras, video projectors, PCs, laptops, and others.

SOURCE

Company originators: Hitachi, Matsushita Electric, Philips, Silicon Image, Sony, Thomson, RCA, and Toshiba.

NATIONAL OR INTERNATIONAL STANDARD

- HDMI Forum Inc.
- Related standard: Consumer Electronics Association/Electronic Industries Alliance (CEA/EIA 861)
- Related standard: Video Electronics Standards Association (VESA) Enhanced Extended Display Identification Data Standard (E-EDID).

KEY FEATURES

- Transfers uncompressed video and compressed or uncompressed audio in digital form between consumer and commercial audio and video equipment.
- The latest version (2.0) supports ultra high-definition 4K video (2160p).
- Up to 32 audio channels.
- Up to 1.536 MHz audio sampling rate for improved fidelity.
- Potential simultaneous delivery of two video streams and four audio streams.

Handbook of Serial Communications Interfaces.
Doi: http://dx.doi.org/10.1016/B978-0-12-800629-0.00043-7

- Support for wide angle 21:9 aspect ratio displays.
- Incorporates High-Bandwidth Digital-Content Protection (HDCP) that prevents recording of video data.
- Provides a 100 Mbps Ethernet channel.

DATA RATE

Gross data rate with 8B/10B encoding per channel. Depends on version. Versions 1.0, 1.1, and 1.2: 1.65 Gb/s. Versions 1.3 and 1.4: 3.4 Gb/s. Version 2.0: 6 Gb/s.

CABLE MEDIUM

Special multiconductor cables with up to 19 wires including four shielded twisted pairs (100 ohms) three for data and one for clock signals, I^2C clock and data lines, control lines and +5V. Two basic types are defined: Standard for connecting 720p and 1080i video and High Speed for connecting 1080p, 4 K, 3D, and deep color video.

RANGE

Typical Standard cable length is 5 m or less. Typical High Speed cable length is up to 15 m. Special extenders and repeaters are available.

CONNECTORS

Five types defined. Type A with 19 pins is the most common. See pin out in Figure 43.1. Type B is a larger connector with 29 pins for more data pairs. Type C is a mini version of A with 19 pins. Type D is a micro version with 19 pins. Type E is a locking version for automotive applications.

NETWORK CONFIGURATION

Point-to-point between devices. See Figure 43.2.

LOGIC LEVELS

Current mode logic (CML) levels: Differential swing from 480 to 1730 mV peak-to-peak.

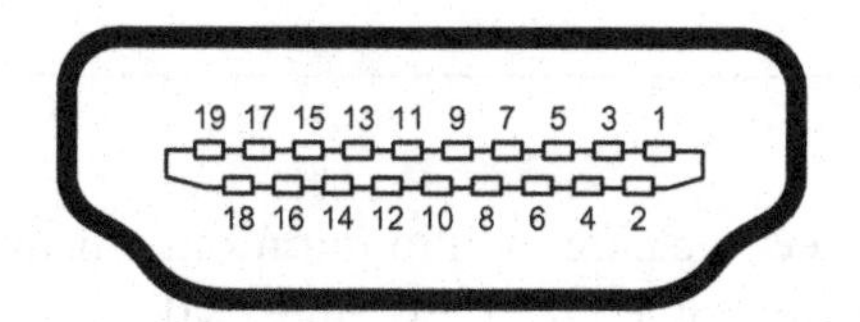

Pin	Signal
1	TMDS Data2+
2	TMDS Data2 Shield
3	TMDS Data2
4	TMDS Data1 +
5	TMDS Data1 Shield
6	TMDS Data1 –
7	TMDS Data0 +
8	TMDS Data0 Shield
9	TMDS Data0 –
10	TMDS Clock+
11	TMDS Clock Shield
12	TMDS Clock–
13	CEC
14	Reserved (N.C. on device)
15	SCL
16	SDA
17	DDC/CEC Ground
18	+ 5V
19	Hot Plug Detect

Figure 43.1 Standard HDMI Type A connector (receptacle) and signal pin designations.

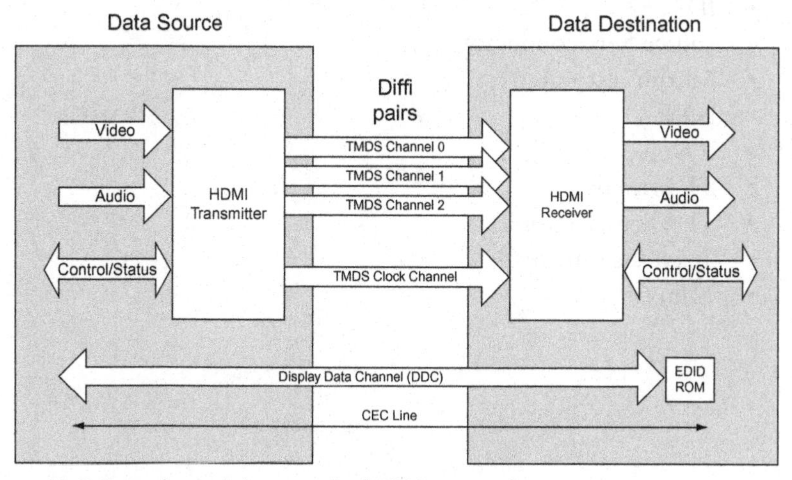

Figure 43.2 Data channels in a standard HDMI connection.

PROTOCOL

There are three separate communications channels (Figure 43.2) each with its own protocol: DDC, CEC, and TMDS.

- Display Data Channel (DDC)
  - Uses I^2C port to read data in E-EDID format.
  - Sending device reads receiving device formats and capability.
- Consumer Electronics Control (CEC)
  - Allows users to control up to 15 devices with one remote control.
  - CEC format lets CEC-enabled devices to control one another.
- Transition Minimized Differential Signaling (TMDS)
  - Time division multiplexed system interleaves audio, video, and control information using standard defined packet formats.
  - Data is transmitted in 10-bit words using a unique 8B/10B code.
  - Packets use 8-bit BCH error correcting codes (ECC).

IC SOURCES

- Analog Devices
- Fairchild
- IDT
- Lattice Semiconductor
- Maxim Integrated
- NXP
- Rohm
- ST Ericsson
- ST Microelectronics
- Texas Instruments
- Xilinx

CHAPTER FORTY-FOUR

HyperTransport (HT)

APPLICATIONS

- Chip-to-chip interface for high speed processors, co-processors, or FPGAs.
- Also useful in routers and switches.

SOURCE

HyperTransport Technology Consortium

NATIONAL OR INTERNATIONAL STANDARD

HyperTransport Technology Consortium

KEY FEATURES

- Flexible serial-parallel bus for chip-to-chip connections of CPUs, etc.
- Competes with PCIe and RapidIO interfaces.
- Scalable data rates.
- Very low latency.
- Four versions: 1.x, 2.0, 3.0, and 3.1.

DATA RATE

- Depends upon number of parallel links and clock rates.
- Clock rates of 800 MHz, 1.4, 2.6, and 3.2 GHz.
- Clocking occurs on both rising and falling edges of the clock signal thereby doubling the data rate.
- With 32-serial lanes and 3.2 GHz clock, maximum aggregate bidirectional data throughput is 51.2 GB/s.

Handbook of Serial Communications Interfaces.
Doi: http://dx.doi.org/10.1016/B978-0-12-800629-0.00044-9

CABLE MEDIUM

PC board traces

RANGE

Inches

NETWORK CONFIGURATION

- Point-to-point only.
- Differential signaling.
- From 2 to 32 serial links possible.
- Two unidirectional links per interface. (e.g. An 8-bit link uses two 8-bit unidirectional (one each for transmit and receive) paths for a total of 16 lanes of two differential connections each).
- Transmit equalization.

LOGIC LEVELS

Similar to LVDS 1.2V.

PROTOCOL

- Packet-based protocol.
- Packets are multiples of 32-bit words.
- Control header is 8 or 12-bytes.
- Address is 40 or 64-bits.
- Data payload is 4 to 64 bytes.
- Transfers are padded to ensure a multiple of 32-bit words.

IC SOURCES

- Advanced Micro Devices (AMD)
- Altera
- Broadcom
- IBM
- Nvidia
- PMC-Sierra
- Xilinx

CHAPTER FORTY-FIVE

InfiniBand (IB)

APPLICATIONS

- Very high-speed interface connections between processors and I/O devices.
- Used in high performance computing (HPC) with super computers.
- Widely used for interconnecting clusters of servers in data centers.
- A common interface for storage area networks (SANs).

SOURCE

InfiniBand Trade Association

NATIONAL OR INTERNATIONAL STANDARD

InfiniBand Trade Association

KEY FEATURES

- Very high speeds.
- Short range.
- Switched fabric interconnections.
- Low latency.
- Maximum BER of 10^{-12}.

DATA RATE

See table for standard rates for a single differential pair. Net rate is less than the gross line rate due to 8B/10B or 64B/66B line encoding to minimize DC build up and to assist in clock recovery.

Handbook of Serial Communications Interfaces.
Doi: http://dx.doi.org/10.1016/B978-0-12-800629-0.00045-0

Speed Name	Gross line rate	Net data rate	Line encoding
Single data rate (SDR)	2.5 Gb/s	2 Gb/s	8B/10B
Double data rate (DDR)	5	4	8B/10B
Quad data rate (QDR)	10	8	8B/10B
Fourteen data rate (FDR)	14.0625	13.5	64B/66B
Enhanced data rate (EDR)	25.78125	24.9	64B/66B
High data rate (HDR)	50	48.5	64B/66B
Next data rate (NDR)	TBD	TBD	TBD

Connections can be aggregated by paralleling four (×4) pairs or twelve (×12) pairs to boost overall transfer rate. Example: ×12 SDR gives 30 Gb/s gross rate and 24 Gb/s net rate.

CABLE MEDIUM

- Dual differential twisted pairs per connection.
- One, four, and twelve pair cables.
- Fiber option.
- Printed circuit board (PCB) connections.
- X4 QDR cable for 40 Gb/s the most common.

RANGE

Depends on cable type and data rate. Copper – SDR: 20 m, DDR: 10 m, QDR: 7 m. Fiber – SDR: 300 m, DDR: 150 m, QDR: 100 or 300 m.

NETWORK CONFIGURATION

- Full duplex.
- Switch fabric architecture. See Figure 45.1.
- Any node to any other node connection possible.
- Multiple topologies: tree, mesh, torus, others.

LOGIC LEVELS

Not defined.

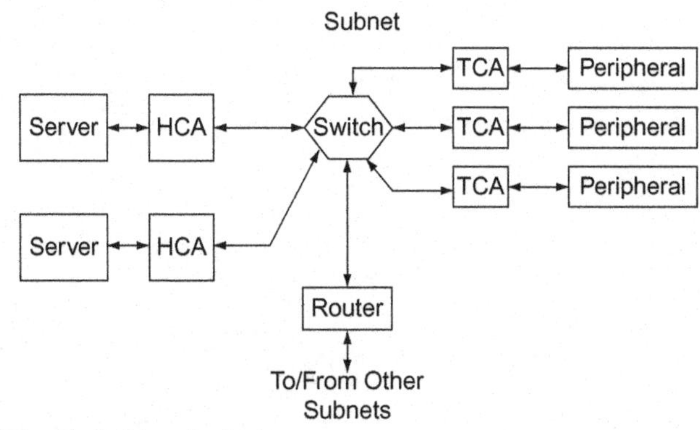

HCA = Host channel adapter connects to CPU.
TCA = Target channel adapter connects to peripheral device.
Switch = Electronic fabric crossbar switch
Router = Interconnects subnetworks.

Figure 45.1 InfiniBand subnet topology and architecture.

PROTOCOL

- Fully defined protocol based on the first four layers of the OSI model.
- Uses IPv6 for addressing at the network layer.
- Two packet frame formats: Link management packets train, direct, and maintain link. Data packets send and receive data, read and write data and acknowledgements.
- Uses both 16-bit and 32-bit CRCs for data integrity.
- Supports remote direct memory access (RDMA).

IC SOURCES

- AppliedMicro
- Broadcom
- Fujitsu
- Hitachi
- Intel
- Mellanox
- QLogic
- Semtech
- Silicon Graphics

CHAPTER FORTY-SIX

JESD204

APPLICATIONS

A serial interface for connecting high speed data converters to DSP, FPGA, or ASIC.

SOURCE

JEDEC Solid State Technology Association

NATIONAL OR INTERNATIONAL STANDARD

JEDEC Solid State Technology Association

KEY FEATURES

- Single lane path for connecting ADCs or DACs to processors or other ICs.
- 8b/10b encoding.
- Optional scrambling.
- Deterministic latency.

DATA RATE

JESD204 and JESD204A: 312.5 Mb/s to 3.125 Gb/s.
JESD204B: Up to 12.5 Gb/s.

CABLE MEDIUM

PC board connections.

Handbook of Serial Communications Interfaces.
Doi: http://dx.doi.org/10.1016/B978-0-12-800629-0.00046-2

RANGE

Up to about 12 inches.

NETWORK CONFIGURATION

- Simplex operation.
- Differential connection.
- Separate clock line shared with converter and source/destination.

LOGIC LEVELS

- Standard can accommodate converter outputs of LVDS, CMOS, or CML.
- JESD204A: 800 mV peak-to-peak
- JESD204B: 360 mV peak-to-peak

PROTOCOL

None

IC SOURCES

- Altera
- Analog Devices
- Lattice
- Linear Technology
- Maxim Integrated
- Texas Instruments
- Xilinx

CHAPTER FORTY-SEVEN

Kandou Bus

APPLICATIONS

Chip-to-chip interconnections in high-speed computers.

SOURCE

Kandou Bus S.A.

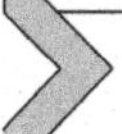

NATIONAL OR INTERNATIONAL STANDARD

None

KEY FEATURES

- Proprietary data encoding/modulation scheme called Chord.
- Potential data speed increases of two to four times current levels.
- Potential power reduction by up to 50%.

DATA RATE

Unspecified. Potential for systems at 56, 100, and 400 Gb/s.

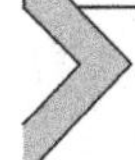

CABLE MEDIUM

PC board interconnections.

RANGE

Up to about 12 inches.

Handbook of Serial Communications Interfaces.
Doi: http://dx.doi.org/10.1016/B978-0-12-800629-0.00047-4

NETWORK CONFIGURATION

- Single-ended connections.
- 2, 4, 6, 8, 9 wire versions.
- Half duplex

LOGIC LEVELS

Standard CMOS or CML/VML.

PROTOCOL

Not available.

IC SOURCES

Kandou Bus S.A.

CHAPTER FORTY-EIGHT

Lightning

APPLICATIONS

Connecting Apple iPhones and iPads to battery chargers or computers.

SOURCE

Apple Inc.

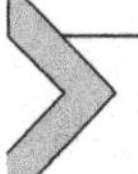

NATIONAL OR INTERNATIONAL STANDARD

Apple Inc.

KEY FEATURES

- 8-pin connector.
- Replaces previous interface with 30-pin connector.
- Connector is reversible.
- Chip in connector dynamically assigns pins to enable reversibility.
- Patented.
- Carries DC power as well as data. Up to 12 watts.

DATA RATE

480 Mbps maximum.

CABLE MEDIUM

Eight-wire cable. See connector and pin assignments in Figure 48.1.

Handbook of Serial Communications Interfaces.
Doi: http://dx.doi.org/10.1016/B978-0-12-800629-0.00048-6

Receptacle View

Pin 1	GND ground
Pin 2	L0p lane 0 positive
Pin 3	L0n lane 0 negative
Pin 4	ID0 identification/control 0
Pin 5	PWR power (charger or battery)
Pin 6	L1n lane 1 negative
Pin 7	L1p lane 1 positive
Pin 8	ID1 identification/control 1

Figure 48.1 The Apple lightning connector with pin outs defined. Note the two differential pairs: L0p/L0n and L1p/L1n.

RANGE

Several feet maximum.

NETWORK CONFIGURATION

- Direct point-to-point connection.
- Two differential data paths. See connector and pin assignment in Figure 48.1.

LOGIC LEVELS

Proprietary.

PROTOCOL

Proprietary.

IC SOURCES

Apple Inc.

CHAPTER FORTY-NINE

Low Voltage Differential Signaling (LVDS)

APPLICATIONS

Physical layer interface for almost any high-speed digital equipment including video displays, wireless infrastructure, storage, and computers.

SOURCE

Originally developed by National Semiconductor (now Texas Instruments).

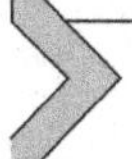

NATIONAL OR INTERNATIONAL STANDARD

TIA/EIA-644 and TIA/EIA-899

KEY FEATURES

- A physical layer-only serial technology.
- High speed (1 to 3 Gb/s).
- Low power consumption.
- Also used as a parallel interface in some applications.

DATA RATE

Depends on length of link but can achieve rates of 1 to 3 Gb/s.

CABLE MEDIUM

Unshielded twisted pair, 100 ohms, PCB traces.

Handbook of Serial Communications Interfaces.
Doi: http://dx.doi.org/10.1016/B978-0-12-800629-0.00049-8

RANGE

Up to 10 m maximum but usually much shorter. Up to 40 m at lower rates with M-LVDS.

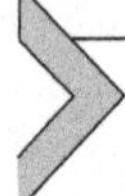

NETWORK CONFIGURATION

- Point-to-point the most common.
- Supports multipoint (M-LVDS) bus configuration with up to 32 nodes.
- Supports half duplex operation.
- Full duplex possible with two cables.
- Uses 8 B/10 B encoding for clock recovery and DC balance.

LOGIC LEVELS

Differential: LVDS – Binary 0: 1V, Binary 1: 1.4V; ±350 mV swing typical with 100-ohm termination. M-LVDS – ±480 mV typical with multiple nodes.

PROTOCOL

None specified. Used as the physical layer for other serial interfaces.

IC SOURCES

- Analog Devices
- IDT
- Maxim Integrated
- Texas Instruments

CHAPTER FIFTY

MIPI Interfaces

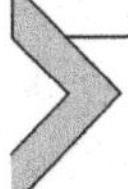

APPLICATIONS

Interfaces to interconnect integrated circuits inside mobile phones.

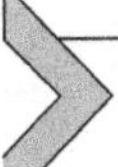

SOURCE

Open Mobile Application Processor Interface (OMAPI) standards established originally by ST Microelectronics and Texas Instruments.

NATIONAL OR INTERNATIONAL STANDARD

Multiple standards maintained by the MIPI Alliance.

KEY FEATURES

- Standardized interfaces between a processor and peripheral chips inside smartphones.
- Individual interfaces for camera, display, RF baseband, multimedia, memory, control, power management, sensors, and test/debug.
- Three physical layer definitions: D-PHY, M-PHY, and C-PHY.

DATA RATE

The minimum and maximum rates per lane.

- M-PHY v3.1: 1.25 to 5.8 Gbps (duplex)
- D-PHY v1.2: 80 Mbps to 2.5 Gbps (simplex)
- C-PHY v1.0: 5.7 Gbps maximum

Handbook of Serial Communications Interfaces.
Doi: http://dx.doi.org/10.1016/B978-0-12-800629-0.00050-4

CABLE MEDIUM

PCB interconnections or short ribbon cable.

RANGE

Inches.

NETWORK CONFIGURATION

- Direct connections, multiple parallel serial lanes.
- One to four differential data lanes.
- Duplex or simplex.
- 8B/10B encoding.

LOGIC LEVELS

Optional peak-to-peak levels:

- M-PHY v3.1: 100–200 mV or 100–400 mV
- D-PHY v1.2: 200 mV or 1.2V
- C-PHY v1.0: 425 mV or 1.3V

PROTOCOL

Multiple protocols. Depends upon application and interface.

IC SOURCES

- ARM
- Intel
- Maxim Integrated
- Nokia
- ON Semiconductor
- Samsung
- ST Microelectronics
- Texas Instruments

CHAPTER FIFTY-ONE

Mobile High-Definition Link (MHL)

APPLICATIONS

Provide a physical connection from a mobile device such as a smartphone or tablet to a high definition video display and/or audio output device. Also used in some PC monitors, A/V receivers, Blu-ray Disc players, or video projectors.

SOURCE

Originally developed by the MHL Consortium made up of companies Nokia, Samsung, Silicon Image, Sony, and Toshiba.

NATIONAL OR INTERNATIONAL STANDARD

MHL Consortium

KEY FEATURES

- Transfers uncompressed high-resolution video and audio from a mobile source to a TV set or other device.
- Four versions: MHL 1, MHL 2, MHL 3, and superMHL. Full backward compatibility.
- Depending upon the version, supports 1080p60, 4Kp30, and 8Kp120 high-definition video formats.
- Provides eight channels of 7.1 surround sound with Dolby TrueHD and DTS-HD.
- Supports up to four or eight displays.
- Remote control capability.
- Uses High-bandwidth Digital Content Protection (HDCP) version 2.2 encryption.
- 5V DC for operation or charging at levels up to 40 watts.
- Easy conversion from HDMI to MHL with an adapter.

Handbook of Serial Communications Interfaces.
Doi: http://dx.doi.org/10.1016/B978-0-12-800629-0.00051-6

DATA RATE

2.25 to 3 Gb/s.

CABLE MEDIUM

Five wire cable with +5V, ground, CBUS control, and a differential data pair.

CONNECTORS

Multiple connector types defined: Micro-USB, USB Type-C, super MHL, HDMI type-A, and some proprietary types. HDMI converters and adapters available.

RANGE

Not specified.

NETWORK CONFIGURATION

- Direct point-to-point connection.
- Differential connection.

LOGIC LEVELS

Not specified.

PROTOCOL

- Uses Transition Minimized Differential Signaling (TMDS) similar to HDMI to time multiplex the video and audio.
- Proprietary protocol.

NOTE: Those unspecified characteristics and detailed specifications indicated above are only available for a fee.

IC SOURCES

- Parade Technologies
- Silicon Image

CHAPTER FIFTY-TWO

Optical Transport Network (OTN)

APPLICATIONS

- Long-range transport of high-speed data over fiber optical networks.
- Internet backbone.

SOURCE

International Telecommunications Union (ITU)

NATIONAL OR INTERNATIONAL STANDARD

ITU G.709 and G.872.

KEY FEATURES

Designed as a "digital wrapper" to carry either synchronous or asynchronous data of popular formats (TCP/IP, SONET, Ethernet).

DATA RATE

- OTU1 – 2.66 Gb/s (To carry SONET OC-48)
- OTU2 – 10.7/11.09 Gb/s (To carry SONET OC-192 or 10 Gigabit Ethernet)
- OTU3 – 43.01 Gb/s (To carry SONET OC-768 or 40 Gigabit Ethernet)
- OTU4 – 112 Gb/s (To carry 100 Gigabit Ethernet)

CABLE MEDIUM

Fiber optical cable, usually single mode fiber.

Handbook of Serial Communications Interfaces.
Doi: http://dx.doi.org/10.1016/B978-0-12-800629-0.00052-8

RANGE

Up to 2000 km.

NETWORK CONFIGURATION

- Point-to-point.
- No formal physical layer defined.

LOGIC LEVELS

IR light levels and Dual Polarization-Quadrature Phase Shift Keying (DP-QPSK).

PROTOCOL

- Packet format.
- See OTN frame or packet in Figure 52.1. Four rows of 4080 bytes (octets) transmitted left to right top to bottom.
- Overhead designates addresses, data type, number of data bytes, and OAMP (operations, administration, maintenance, and provisioning) data for management, and troubleshooting.
- Payload contains the IP, Ethernet, or SONET frames to be transmitted.
- Forward error correction (FEC) is a variation of Reed Solomon designated RS (255, 239) that provides 6.2 dB of coding gain that extends the reach.
- BER is 10^{-12}.

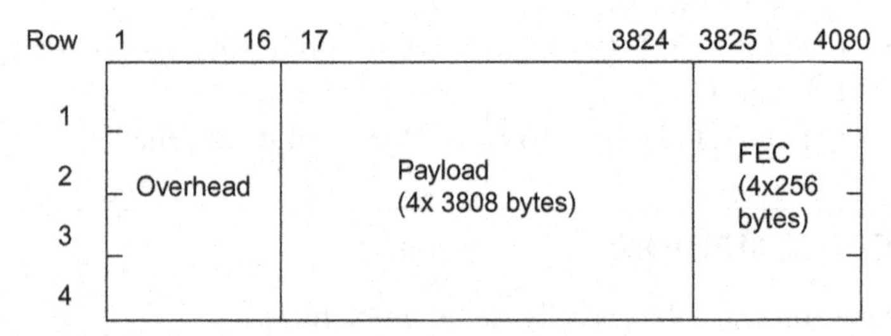

Figure 52.1 Simplified OTN frame.

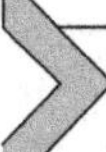

IC SOURCES

- Altera
- Applied Micro
- Cortina
- IP Light
- PMC-Sierra
- TPACK
- Vitesse
- Xilinx

CHAPTER FIFTY-THREE

PCI Express (PCIe)

APPLICATIONS

- The interface between the processor and most peripheral chips and devices in modern personal computers, laptops, and servers.
- Used on motherboard connectors, backplanes, and expansion cards.

SOURCE

Original source is Intel.

NATIONAL OR INTERNATIONAL STANDARD

PCI Special Interest Group (PCI-SIG)

KEY FEATURES

- Replaces previous parallel PCI and PCI-X parallel buses whose speed capability topped out because of losses, cross talk, and data skew.
- Serial data paths significantly boost data speeds between processor and peripherals.
- Multiple speed versions available to achieve desired rate with minimum hardware and cost.
- Supplies DC power.
- Hot pluggable.
- Data transfer via full duplex lanes using two differential pairs (×1).
- Multiple lanes are used to form links that increase overall throughput (×4, ×8, ×16, ×32). For example, a link of ×4 uses four lanes.

Handbook of Serial Communications Interfaces.
Doi: http://dx.doi.org/10.1016/B978-0-12-800629-0.00053-X

DATA RATE

- Lane speed depends upon PCIe version.
  - Ver. 1 – 2.5 Gbps with 8B/10B encoding.
  - Ver. 2 – 5.0 Gbps with 8B/10B encoding.
  - Ver. 3 – 8.0 Gbps with 128B/130B encoding.
  - Ver. 4 – 16 Gbps with 128B/130B encoding.
- Net bandwidth with encoding.
  - Ver. 1 – 250 MB/s with 8B/10B encoding.
  - Ver. 2 – 500 MB/s with 8B/10B encoding.
  - Ver. 3 – 984.6 MB/s with 128B/130B encoding.
  - Ver. 4 – 1969.2 MB/s with 128B/130B encoding.

CABLE MEDIUM

Printed circuit board traces. Short cables, length undefined.

CONNECTORS

Special connectors designed for different speed levels to be used on motherboards and plug-in cards.

RANGE

Inches.

NETWORK CONFIGURATION

- Point-to-point connection between CPU and end device.
- High speed switches are used to allow CPU to serve multiple peripherals.
- Multiple parallel serial lanes increase overall throughput. Byte interleaving used.

LOGIC LEVELS

Low-Voltage Differential Signaling (LVDS): ±300 mV.

PROTOCOL

- Three-layer OSI-like protocol using the physical layer, data link layer, and a transaction layer.
- Data transfer by packets.
    - Data payload from software (up to 1024 32-bit words) is encapsulated in a transaction layer packet with ID header and a 32-bit CRC.
    - Data link layer encapsulates transaction packet with a sequence header and another 32-bit CRC.
    - Physical layer adds start and end fields to packet.

IC SOURCES

- AMD
- Avago
- Intel
- Nvidia
- NXP
- PLX Technology

CHAPTER FIFTY-FOUR

Passive Optical Networks (PON)

APPLICATIONS

- Broadband connectivity.
- Internet service.
- Voice over Internet Protocol (VoIP)
- Video services.

SOURCE

Telecommunications and Internet companies.

NATIONAL OR INTERNATIONAL STANDARD

- IEEE 802.3ah, 802.3av.
- ITU G.983, G.984, G.987.

KEY FEATURES

- Designed for local metropolitan area networks (MANs).
- Uses only passive combiners and splitters with no electrical-to-optical or optical-to-electrical (OEO) conversions for repeaters.
- Enables lower cost optical networks.
- Multiple versions and standards.
- Uses wavelength division multiplexing of downstream (DS) and upstream (US) data traffic on different IR wavelengths.

DATA RATE

Depends on the version and standard. Typical maximums:

- APON (ATM PON) 622 Mb/s downstream (DS), 155 Mb/s upstream (US).

Handbook of Serial Communications Interfaces.
Doi: http://dx.doi.org/10.1016/B978-0-12-800629-0.00054-1

- BPON (Broadband PON) 622 Mb/s downstream (DS), 155 Mb/s upstream (US).
- GPON (Gigabit PON) 2.488 Mb/s DS, 1.244 Mb/s US
- XGPON (10 Gigabit PON) 10 Gb/s DS, 2.5 Gb/s US
- EPON 1 Gb/s DS and US
- 10G-EPON 10 Gb/s DS, 1 Gb/s US

CABLE MEDIUM

Fiber optic cable.

RANGE

20 km maximum.

NETWORK CONFIGURATION

- Multi-branch tree. See Figure 54.1.
  - OLT is the optical line terminal the main data source.
  - ONU/ONT are the optical network units or optical networking terminals are the customers node equipment.

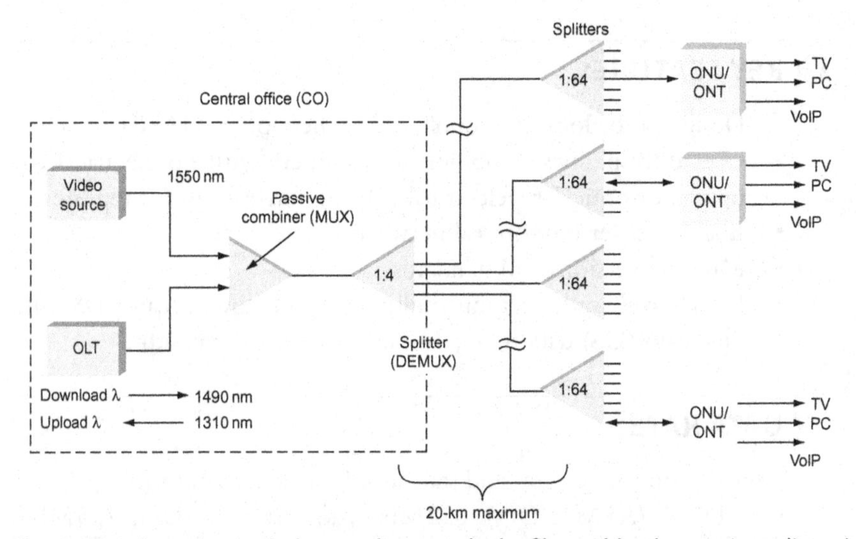

Figure 54.1 A passive optical network uses only the fiber cable plus passive splitters/combiners and no electrical regeneration.

- Downstream data is broadcast to all nodes.
- Upstream data from customers is time multiplexed. (TDMA)

LOGIC LEVELS

IR light levels

PROTOCOL

- APON, BPON, GPON, and XGPON use the ATM protocol.
- EPON uses the standard Ethernet protocol.

IC SOURCES

- AMCC
- Broadcom
- Broadlight
- Centillium
- Cortina
- Freescale
- Passave'
- PMC-Sierra
- Teknovus

CHAPTER FIFTY-FIVE

RapidIO

APPLICATIONS

Equipment used in high-performance computing (HPC), data center servers, networking, storage, cellular communications infrastructure, industrial controls, aerospace, and military.

SOURCE

Originally developed by Mercury Computer Systems and Motorola (now Freescale/NXP).

NATIONAL OR INTERNATIONAL STANDARD

RapidIO Trade Association.

KEY FEATURES

- Chip-to-chip connections between multiple processors, DSPs and peripheral devices or memory and I/O.
- PCB-to-PCB connections via a backplane.
- Chassis-to-chassis connections with very short cables.
- Very high-speed multiple parallel serial links.
- Uses a switch fabric architecture.
- Competes with Ethernet and PCIe.
- Low latency.

DATA RATE

Varies with version and number of serial lanes. The standard defines 1, 2, 4, 8, and 16 serial lanes with speeds of 1.25, 2.5, 3.125, 5, 6.25, and 10 Gb/s data rates. For example, 4× lanes at 6.25 Gb/s provides a 25 Gb/s net throughput. Total data rate to 160 Gb/s.

Handbook of Serial Communications Interfaces.
Doi: http://dx.doi.org/10.1016/B978-0-12-800629-0.00055-3

CABLE MEDIUM

PCB copper connections. Short cables.

RANGE

Less than 1 m.

NETWORK CONFIGURATION

- Switch fabric between multiple processors, memory, and I/O. See Figure 55.1.
- 2–1000+ possible nodes.
- Full duplex.
- Uses the Ethernet XAUI physical layer specifications for 1.25, 2.5, and 3.125 GBaud rates.
- Uses OIF CEI 6+ specifications for 5 and 6.25 GBaud rates.
- Uses Ethernet 10GBASE-KR specifications for 10.3125 GBaud.
- Uses 8B/10B coding for speeds to 6.25 GBaud.
- Uses 64B/67B coding for data rates over 6.25 GBaud.

LOGIC LEVELS

See logic levels for XAUI, OIF CEI.

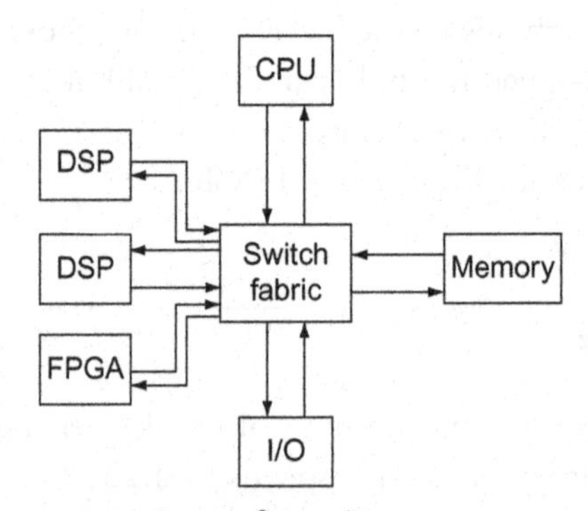

Figure 55.1 RapidIO representative configuration.

PROTOCOL

- Uses layers similar to the OSI model.
  - Layer 1: Logical – Request and response transactions.
  - Layer 2: Transport – Source and destination IDs.
  - Layer 3: Physical – 1, 2, 4, 8, or 16 lanes of serial data.
- Uses one or more 32-bit packets for control and data.
- Data payload up to 256 bytes.

IC SOURCES

- Altera
- Freescale/NXP
- IDT
- Mercury
- Texas Instruments
- Xilinx

CHAPTER FIFTY-SIX

Serial Attached SCSI (SAS)

APPLICATIONS

- Replaces older original parallel Small Systems Computer Interface (SCSI).
- Connects computers to storage units like hard disk, optical, tape, or solid-state drives.

SOURCE

- American National Standards Institute.
- International Committee for Information Technology Standards (T10 technical committee).
- SCSI Trade Association.

NATIONAL OR INTERNATIONAL STANDARD

ANSI/INCITS 376-2003.

KEY FEATURES

- A form of direct attached storage (DAS).
- Extends transmission range and data speed of older parallel SCSI.
- Simpler and lower cost connections.
- Similar in function and operation to Serial Advanced Technology Attachment (SATA).

DATA RATE

3, 6, and 12 Gb/s.

Handbook of Serial Communications Interfaces.
Doi: http://dx.doi.org/10.1016/B978-0-12-800629-0.00056-5

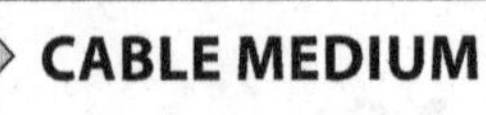

CABLE MEDIUM

Multiwire cable.

CONNECTORS

Multiple connectors defined with 26, 29, 32, and 36 pins for connections to 1, 2, 4, or 8 devices.

RANGE

Up to 10 m.

NETWORK CONFIGURATION

- Point-to-point connections between initiators (PCs, servers) and targets (disk drives).
- Expanders are switches that are used to increase the number of targets on one initiator port.

LOGIC LEVELS

Differential signaling, 800–1600 mV peak-to-peak.

PROTOCOL

- Multilayer: Physical, Link, Port, Transport, and Application.
- 8b/10b encoding.

IC SOURCES

- Avago/LSI
- Intel
- Marvell
- PMC-Sierra
- Vitesse

CHAPTER FIFTY-SEVEN

Serial Advanced Technology Attachment (SATA)

APPLICATIONS

Connects storage units like hard disks and optical disks to a PC or server. Used in most modern PCs and laptops.

SOURCE

Original source IBM and Western Digital who developed the Integrated Device Electronics (IDE) interface for the original IBM PC/AT now called parallel ATA.

NATIONAL OR INTERNATIONAL STANDARD

- International Committee for Information Technology Standards (T13 technical subcommittee).
- Serial ATA International Organization (SATA-IO).

KEY FEATURES

- A form of internal direct attached storage (DAS).
- Extends transmission range and data speed of older parallel ATA.
- Simpler and lower cost cables and connections.
- Similar in function and operation to Serial Attached SCSI (SAS).
- Hot pluggable.
- The eSATA version is designed to connect to external drives.

DATA RATE

- SATA ver.1 1.5 Gb/s
- SATA ver.2 3.0 Gb/s

Handbook of Serial Communications Interfaces.
Doi: http://dx.doi.org/10.1016/B978-0-12-800629-0.00057-7

- SATA ver.3 6.0 Gb/s
- SATA ver. 3.2 (Express) 16 Gb/s

CABLE MEDIUM

- Seven-wire data cable and connector with two differential data pairs and three grounds.
- Separate 15-wire cable and connector for DC power (3.3, 5, and 12V).
- Different seven-wire shielded cable for eSATA.

RANGE

- Maximum 1 m.
- Maximum 2 m for eSATA.

NETWORK CONFIGURATION

- Direct point-to-point connection for each drive.
- A Multipler device serves as a hub to permit one PC port to support up to 15 storage devices.

LOGIC LEVELS

- LVDS, AC coupled.
- Differential, 400–600 mV peak-to-peak typical.
- 500–600 mV for eSATA version.

PROTOCOL

- Multilayer: Physical, Link, and Transport layers.
- 8b/10b encoding.

IC SOURCES

- Advanced Micro Devices (AMD)
- Intel
- Marvell
- NEC

CHAPTER FIFTY-EIGHT

Serial Digital Interface (SDI)

APPLICATIONS

Connect video equipment in television facilities and professional studios.

SOURCE

Society of Motion Picture and Television Engineers (SMPTE).

NATIONAL OR INTERNATIONAL STANDARD

SMPTE 259M, 292M, 344M, 372M, 424M, ST-2081, ST-2082.

KEY FEATURES

- Transmission of uncompressed and unencrypted video.
- Also transmits digital audio and time codes.
- Multiple versions to handle a wide range of video formats and resolutions including 480i, 480p, 576i, 720p, 1080i, 1080p, 4Kp30, 4Kp60.

DATA RATE

- SD-SDI: 143, 177, 270, and 360 Mb/s
- ED-SDI: 540 Mb/s
- HD-SDI: 1.485 Gb/s
- HD-SDI (Dual Link): 2.97 Gb/s
- 3G-SDI: 2.97 Gb/s
- 6G-UHD-SDI: 6 Gb/s
- 12G-UHD-SDI: 12 Gb/s
- 24G-UHD-SDI: 24 Gb/s

Handbook of Serial Communications Interfaces.
Doi: http://dx.doi.org/10.1016/B978-0-12-800629-0.00058-9

CABLE MEDIUM

75-Ω coax cable with BNC connectors.

RANGE

Depends upon data rate. Standard SDI at 270 Mb/s can achieve up to 300 m. HD-SDI versions less than 100 m.

NETWORK CONFIGURATION

- Point-to-point connection.
- Simplex.
- Single-ended/unbalanced.

LOGIC LEVELS

- 800 mV peak-to-peak.
- NRZI encoding.

PROTOCOL

- Proprietary protocol.
- 10-bit or 20-bit data packets.
- Synchronization packets.
- Line numbering packets.
- CRC.
- Ancillary packets for audio, closed captioning, and time codes.

IC SOURCES

- Altera
- Extron
- Gennum
- M/A-COM/Mindspeed
- Texas Instruments
- Xilinx

CHAPTER FIFTY-NINE

Synchronous Optical Network (SONET) and Synchronous Digital Hierarchy (SDH)

APPLICATIONS

- Original application to carry digital telephony between central offices and other systems.
- Metropolitan area networks (MAN) and wide area networks (WAN).
- Internet connectivity.
- SONET is used in the United States and Canada while SDH is used in Europe and the rest of the world.

SOURCE

Original source Telecordia Technologies (now part of Ericsson).

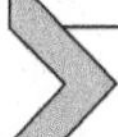

NATIONAL OR INTERNATIONAL STANDARD

- Telcordia: GR-253, GR-499.
- American National Standards Institute (ANSI): SONET T1.105, T1.119.
- International Telecommunications Union – Telecommunications Standards Sector (ITU-T): SDH G.707, G.783, G.803.

KEY FEATURES

- High-speed fiber optical network.
- Data speeds to 40 Gb/s.
- Long-range connections for MAN and WAN.
- High reliability.

Handbook of Serial Communications Interfaces.
Doi: http://dx.doi.org/10.1016/B978-0-12-800629-0.00059-0

- Ability to carry asynchronous data such as ATM, Ethernet, and TCP/IP.
- Now gradually being replaced by OTN.

DATA RATE

SONET level	STM level	Data (line) rate
STS-1/OC-1	–	51.84 Mb/s
STS-3/OC-3	STM-1	155.52 Mb/s
STS-12/OC-12	STM-4	622.08 Mb/s
STS-48/OC-48	STM-16	2.488 Gb/s
STS-192/OC-192	STM-64	9.953 Gb/s
STS-768/OC-768	STM-256	39.812 Gb/s

Basic transmission framing units:

- STS = Synchronous Transport Signal (US version)
- OC = Optical Carrier
- STM = Synchronous Transport Module (European version)

CABLE MEDIUM

Fiber optical cable, typically single-mode fiber (SMF).

RANGE

Depends upon optical system. Many kilometers.

NETWORK CONFIGURATION

- Point-to-point.
- Ring (unidirectional or bidirectional) with add-drop multiplexers (ADMs) nodes. See Figure 59.1.

NOTE: ADMs allow data to be added or extracted from the data stream at any time at the desired data rate.

LOGIC LEVELS

IR light levels.

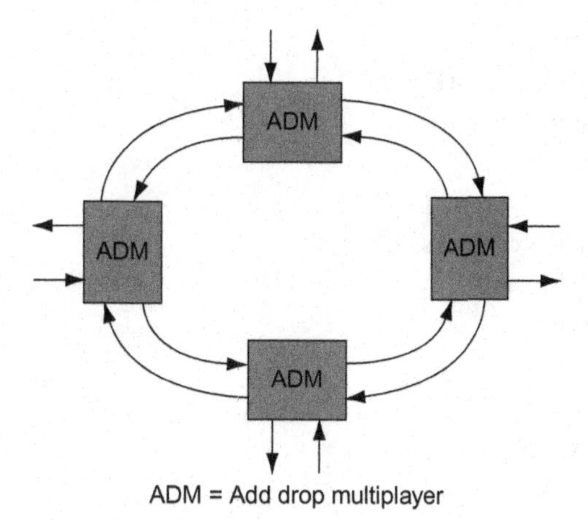

Figure 59.1 SONET ring with ADM nodes.

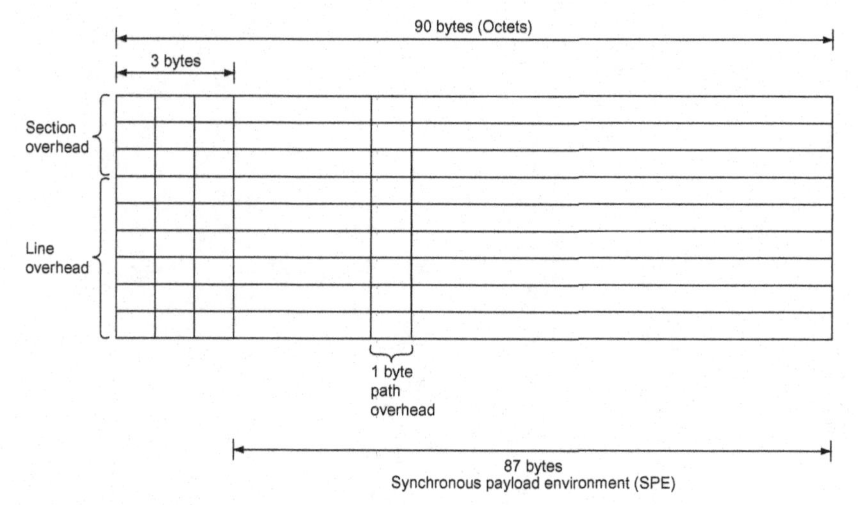

Figure 59.2 Basic SONET frame format.

PROTOCOL

- Time division multiplexed data with overhead.
- Basic SONET frame is shown in Figure 59.2.

IC SOURCES

- Altera
- Applied Micro

- Cypress
- Galazar Networks
- Infineon
- Inphi
- Intel
- Parama Networks
- PMC-Sierra
- TPACK
- TranSwitch
- Vitesse
- Xilinx

CHAPTER SIXTY

Thunderbolt

APPLICATIONS

- Connecting high-speed PCs or laptops to one another for data transfer.
- Connect high-speed peripherals such as storage devices, high-resolution video monitors, or cameras to PCs or laptops.
- Used primarily on Apple Mac PCs and laptops.

SOURCE

Developed jointly by Apple and Intel.

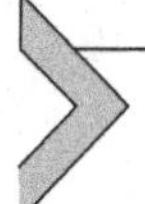

NATIONAL OR INTERNATIONAL STANDARD

Intel.

KEY FEATURES

- Original code name Light Peak.
- Combines the PCI Express and DisplayPort interfaces on a single cable/interface.
- Hot pluggable.
- Supplies DC power externally up to 10W.
- Two complete full duplex lanes.

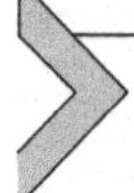

DATA RATE

10 Gb/s bidirectional on two channels.

Handbook of Serial Communications Interfaces.
Doi: http://dx.doi.org/10.1016/B978-0-12-800629-0.00060-7

CABLE MEDIUM

- Twenty wire active copper cable.
- Optional fiber optical cable.

RANGE

- 3 m (copper)
- 50 m (optical)

NETWORK CONFIGURATION

- Point-to-point. See Figure 60.1.
- Daisy chain up to six devices.

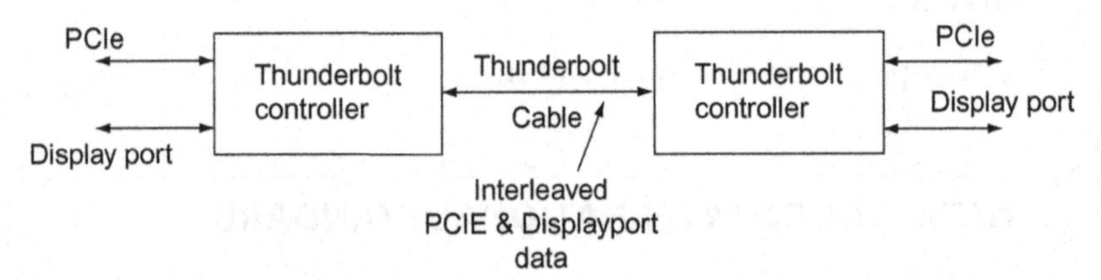

Figure 60.1 Thunderbolt interface concept.

LOGIC LEVELS

Same as PCIe and DisplayPort.

PROTOCOL

Uses PCIe and DisplayPort protocols.

IC SOURCES

- Gennum
- Intel
- Pericom
- Texas Instruments

CHAPTER SIXTY-ONE

Ten and Forty Gigabit Ethernet Attachment Unit Interface (XAUI and XLAUI)

APPLICATIONS

A chip-to-chip or chip-to-module interface between Ethernet Media Access Controller (MAC) layer 2 circuitry and the physical layer 1 module used by the main Ethernet medium (fiber optic cable).

SOURCE

Institute of Electrical and Electronic Engineers (IEEE).

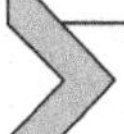

NATIONAL OR INTERNATIONAL STANDARD

IEEE 802.3ba and extensions.

KEY FEATURES

- An interface that reduces the number of data connections between chips and modules in an Ethernet system.
- Reduces the 10 Gb Media Independent Interface (XGMII) that features two 32-bit data buses plus two 4-bit control buses to four 3.125 Gb serial interfaces.
- Another version (XLAUI) reduces the 10 Gb Media Independent Interface (XGMII) that features two 32-bit data buses plus two 4-bit control buses to four 10 Gb serial interfaces.

Handbook of Serial Communications Interfaces.
Doi: http://dx.doi.org/10.1016/B978-0-12-800629-0.00061-9

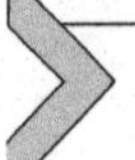

DATA RATE

40 or 100 Gb/s.

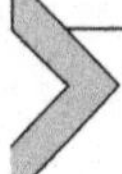

CABLE MEDIUM

PCB traces. Back planes.

RANGE

Up to 50 cm.

NETWORK CONFIGURATION

- The interface circuitry between the XGMII and the XAUI is called the 10 Gb Ethernet Extended Sublayer (XGXS). See Figure 61.1.
- XAUI – four transmit and four receive lanes at 3.125 Gb/s.
- XLAUI – four transmit and four receive lanes at 10.3125 Gb/s.
- 8b/10b encoding on XAUI.
- 64b/66b encoding on XLAUI.

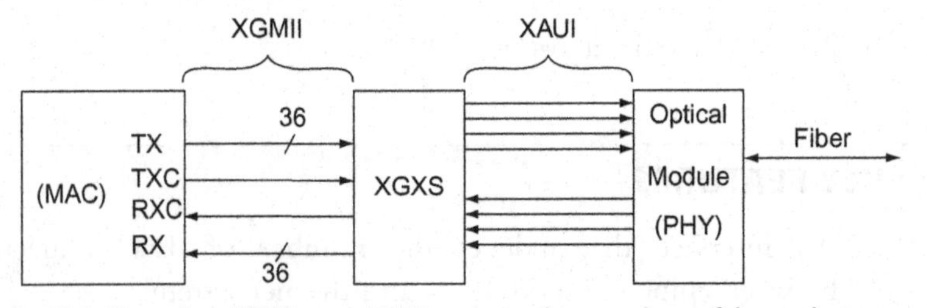

Figure 61.1 How the XGXS and XAUI reduce the number of lanes between the Ethernet MAC and PHY.

LOGIC LEVELS

Binary 0: −0.3V, Binary 1: +0.3V.

PROTOCOL

- A physical layer standard only.
- Uses the standard Ethernet protocol and some related Media Independent Interface (MII) at the MAC layer.

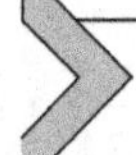

IC SOURCES

- Altera
- Lattice
- Xilinx

CHAPTER SIXTY-TWO

Testing Considerations

Practically all electronic devices and products have a serial interface. It is difficult to name a product that does not use one. For that reason, at some time in your engineering career you will most likely have to test or troubleshoot a serial interface. Most interface standards contain measurement specifications and many even recommend specific test procedures. Here are the key factors to consider in such testing.

DOCUMENTATION

As a first step, you should acquire the actual serial standard documentation. Go to the standard agency or sponsoring organization and order the detailed standard materials with all the addenda and updates. There is usually a fee involved but it is worth it. You cannot test a complex interface without knowing the details. Furthermore, you want your product to have full interoperability with other devices using the standard. Most documentation also includes detailed test procedures to help you set up your tests. This will save time and ensure your procedures are correct.

DATA RATE

Data rate is measured with an oscilloscope. The oscilloscope shows the bit time interval (t) that is used to compute the actual bit frequency (f).

$$f = 1/t$$

BIT ERROR RATE

Bit error rate (BER) is a measure of the number of bit errors that occur in a given number of bit transmissions. It is usually expressed as a ratio. For example, if 5 bit errors occur in one million bits transferred, the BER is 5/1,000,000 or 5×10^{-6}. BER is a measure of the quality of the transmitting device, the receiver, the transmission path and its environment as it takes

Handbook of Serial Communications Interfaces.
Doi: http://dx.doi.org/10.1016/B978-0-12-800629-0.00062-0

into consideration factors such as noise, jitter, attenuation, fading, and any error detection and correction schemes used in the interface standard.

BER is measured by applying a pseudorandom continuous NRZ bit stream to the interface, counting the bit errors and comparing the transmitted to the received data then computing the ratio. BER testers (BERTs) are available to automate this measurement for you.

BER testing is generally not used with low-speed interfaces. It is an essential test for high-speed interfaces.

JITTER

Jitter is the rapid time shift of the rising and falling edges of a pulse signal. It is a kind of low deviation frequency modulation of a data signal. Jitter is caused by clock instability, PLL phase noise, and other noise sources. Jitter causes bit errors and other timing problems.

Jitter is measured by actually measuring individual repeating signal periods (>1000) then averaging them to get root mean square (rms) jitter. This is called cycle to cycle jitter. You can also calculate the standard deviation of the signal cycles to get another measure of jitter. Long term or accumulated jitter is jitter measured over a larger number of signal cycles such as 10,000. It is often a better measure of jitter performance than the cycle to cycle jitter measurement described above.

Jitter is usually indicated as the average peak-to-peak time interval. The unit of measurement is usually in picoseconds (ps). Jitter is generally best measured with statistics because of its random characteristic.

Jitter is mainly a factor at the higher data rates (100 Mb/s or more) where the rise and fall times become a significant percentage of the bit time. It is rarely an issue in the lower data rate interfaces.

EYE DIAGRAMS

An eye diagram or eye pattern is simply a graphical display of a serial data signal with respect to time that shows a pattern that resembles an eye. See Figures 62.1 and 62.2. The signal at the receiving end of the serial link is connected to an oscilloscope and the sweep rate is set so that one or two bit time periods (unit intervals or UI) are displayed. This causes bit periods to overlap and the eye pattern to form around the upper and lower signal levels and the rise and fall times. The eye pattern readily

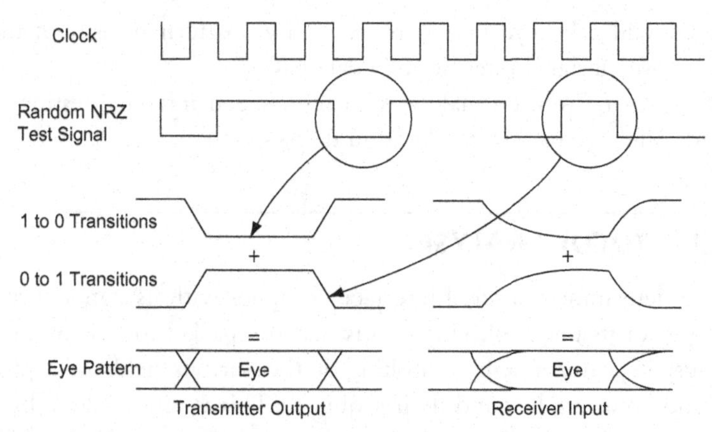

Figure 62.1 The eye pattern is formed by overlapping sweeps and shows all signal aberrations.

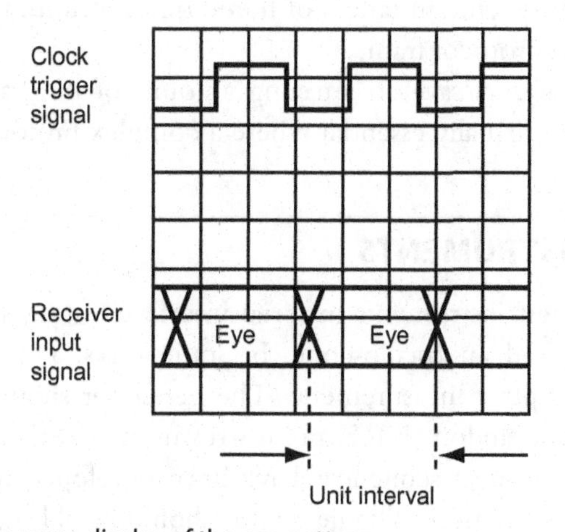

Figure 62.2 Oscilloscope display of the eye pattern.

shows the rise and fall time lengthening and rounding as well as the horizontal jitter variation.

When displaying a differential signal, two scope channels are used, one on each side of the receiver input. These two signals, measured with respect to ground, are then algebraically summed (subtracted) in the scope to provide the true display.

The eye pattern is a composite signal that reflects the channel bandwidth, attenuation, jitter, and rise/fall time variations. As the signal frequency increases for a given bandwidth and the signal is attenuated and interfered with, the eye begins to close. Eye pattern measurements can show the overall signal integrity of a data path.

Eye patterns are used mainly with high-speed interface testing. Some standards offer eye pattern masks as guidelines.

PROTOCOL ANALYSIS

To determine if a serial interface complies with its standard requires an analysis of its protocol. This means that the packet format of an interface must be determined by looking at the various fields of a protocol frame and counting bits and their sequence. This is usually done by using a protocol analyzer, packet analyzer, or network analyzer. A protocol analyzer is a hardware device, a combination of hardware and software, or embedded software inside a sampling oscilloscope. The analyzer displays the bit details of specific instances of frame transmissions. It can determine the veracity of a protocol frame.

Protocol analyzers save an amazing amount of time and simplify the testing. They are virtually essential where a complex protocol is used.

TEST INSTRUMENTS

The key test instruments in serial interface testing are a digital pattern generator and an oscilloscope. In some cases, a spectrum analyzer can be used for jitter measurements. The generator should be capable of producing pseudorandom NRZ pulse streams at the desired data rates. A wide range of test data sequences have been developed to test interfaces under different conditions. The generator should be able to generate these test sequences directly or by external programming. In some instances, the testing oscilloscope software may provide the test data stream.

The oscilloscope is the main test instrument for serial data testing. Most oscilloscope manufacturers provide test software options for the most common serial interface standards. In some cases, special test instruments are available to measure specific characteristics such as jitter or BER. Check for the availability of specific test hardware and software with the oscilloscope manufacturers including Anritsu, Keysight Technologies, National Instruments, Rhode & Schwarz, Tektronix, and Teledyne Lecroy. Companies like EXFO and JDSU also provide test instruments for fiber optic interfaces.

SECTION IV

Broadband Interfaces

CHAPTER SIXTY-THREE

Broadband Interface Concepts

The interfaces covered in the previous chapters put the serial binary voltages directly on the cable. This approach is known as baseband operation. Another method of transmitting serial data is to use broadband methods that employ modulation. The serial binary data is used to modulate a higher frequency carrier signal that is then transmitted over the cable. This chapter is a general introduction to the modulation methods used. Chapters summarizing the most widely used broadband interfaces follow.

THE RATIONALE FOR MODULATION

Applying the high-speed binary signals directly to the cable causes the signals to be greatly influenced by the characteristics of the cable. The cable is a transmission line that is in effect a long low-pass filter. The cable wires introduce distributed series inductance and resistance as well as shunt capacitance. These attenuate and distort the signals thereby limiting the distance the signal can be faithfully transmitted. The introduction of noise is another limiting factor.

One way to extend the transmission range is to have the binary data modulate a higher frequency signal called a carrier. This binary signal does not experience as much distortion this way, however, the carrier is affected with attenuation and noise. This approach allows much longer cable runs and in some forms permits higher data rates than those possible with baseband techniques.

There are basic modulation methods: amplitude modulation (AM), frequency modulation (FM), and phase modulation (PM). These methods state how the sine wave carrier is modified by the binary data. There are multiple versions of these methods for transmitting binary data. The following shows the modulation concept with analog signals.

Handbook of Serial Communications Interfaces.
Doi: http://dx.doi.org/10.1016/B978-0-12-800629-0.00063-2

AMPLITUDE MODULATION

In amplitude modulation, it is the voltage level of the signal to be transmitted that changes the amplitude of the carrier in proportion, see Figure 63.1. With no modulation, the AM carrier is transmitted by itself. When the modulating information signal (a sine wave) is applied, the carrier amplitude rises and falls in accordance. The carrier frequency remains constant during amplitude modulation.

Analog amplitude modulation is widely used in radio. AM broadcast stations are, of course, amplitude modulated. So are citizen's band radios and aircraft radios. A special form of amplitude modulation, known as quadrature amplitude modulation (QAM), is also widely used in modems to transmit digital data over cable or wireless, more on that discussed later.

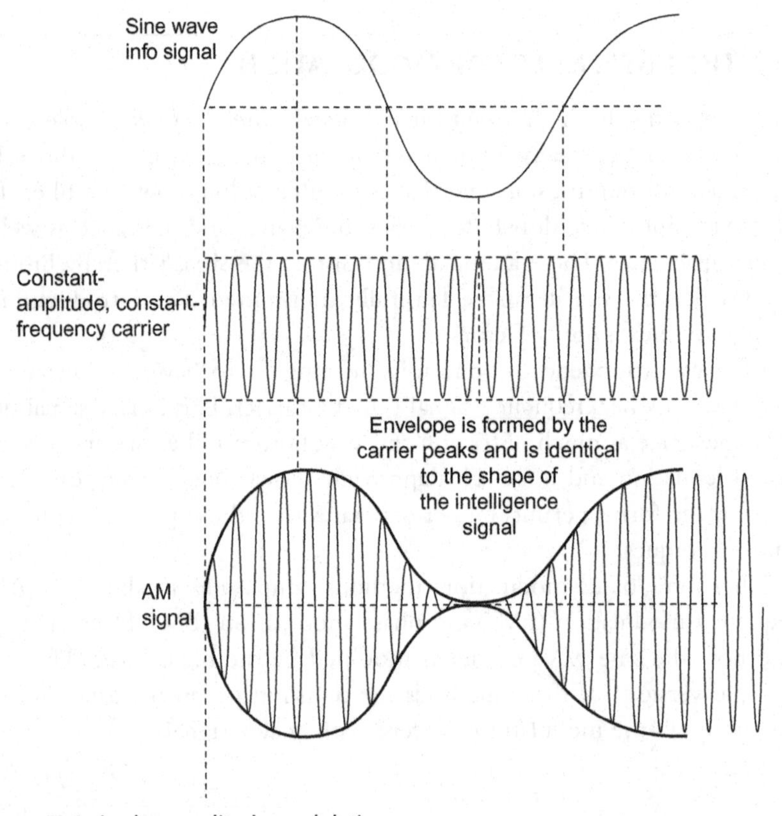

Figure 63.1 Analog amplitude modulation.

SIDEBANDS

The modulation process causes new signals to be generated. These new sine wave signals are called sidebands. Their frequencies are the sum and difference of the carrier and modulating signal frequencies. For example, in an AM radio station, audio frequencies as high as 5 kHz can be transmitted. If a 5 kHz sine wave tone is to be transmitted, the modulation process causes sidebands 5 kHz below and 5 kHz above the carrier to be produced. This is illustrated in Figure 63.2. For an AM radio station with a carrier frequency of 860 kHz, the lower sideband (LSB) would occur at 860 − 5 = 855 kHz while the upper sideband (USB) occurs at 860 + 5 = 865 kHz. The carrier and the sidebands combined produce the composite waveform shown at the bottom of Figure 63.1.

Since binary signals are rectangular waves they contain many odd and even harmonics according to the Fourier theory. Each of the harmonics will also generate a sideband above and below the carrier.

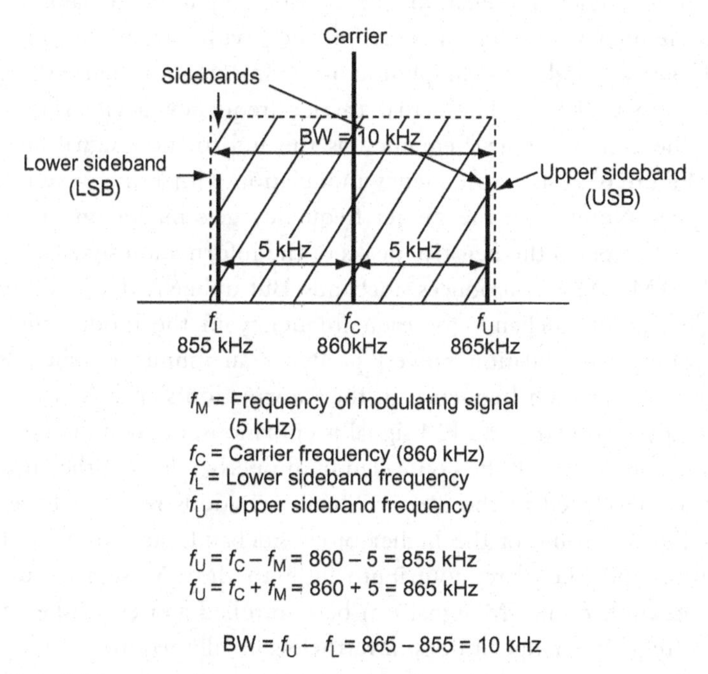

Figure 63.2 Sidebands and bandwidth for AM radio.

BANDWIDTH

This brings up the very important concept of bandwidth. Bandwidth refers to a range of frequencies that the modulated signal occupies. This is referred to as the channel. As you can see from Figure 63.2, the AM signal consists of the carrier and the sidebands. Together these signals occupy a bandwidth of 10 kHz. You can figure out the bandwidth of any signal by simply subtracting the LSB frequency from the USB frequency.

$$BW = USB - LSB = 865 - 855 = 10\,\text{kHz}$$

All of the transmitting and receiving circuits must pass signals in this 10 kHz range in order to avoid distortion of the signal or lost information. Binary signals will, of course, produce many more sidebands and much wider bandwidth signals requiring a wider channel.

FREQUENCY MODULATION

In frequency modulation, the carrier amplitude remains constant but its frequency is changed in accordance with the modulating signal. Specifically, the higher the amplitude of the information signal, the greater the frequency change. The actual carrier frequency deviates above and below the center carrier frequency as the information signal amplitude varies. Figure 63.3 shows frequency modulation with a sine wave information signal. Notice that the carrier frequency gets higher on the positive peaks and lower on the negative peaks of the information signal.

Like AM, FM also produces sidebands. But unlike AM which produces a single pair of sidebands for each frequency in the modulating signal, the frequency modulation process produces an infinite number of pairs of sidebands for each frequency in the information signal. As a result, the bandwidth occupied by an FM signal is enormous. Luckily, the number of sidebands produced can be controlled by properly selecting the amount of deviation permitted in the carrier. Small deviations result in fewer sidebands. Further, some of the higher order sidebands are extremely low in amplitude and, therefore, contribute little to the FM signal. But while the bandwidth of an FM signal can be controlled and established to fit a desired frequency range, it does nevertheless usually require a wider bandwidth channel than an AM signal.

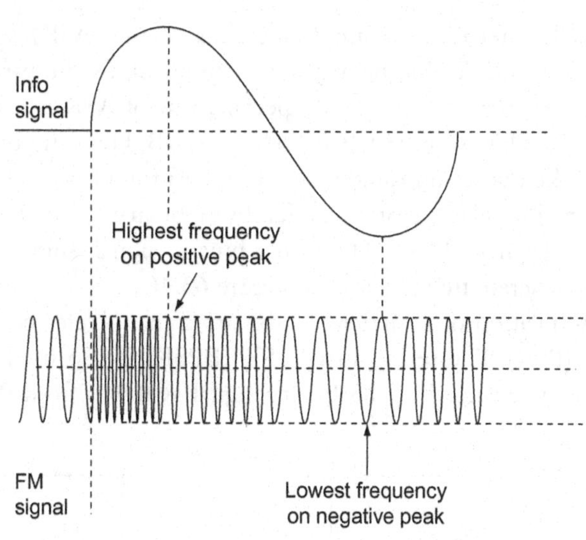

Figure 63.3 Analog frequency modulation.

The primary benefit of FM is that it is less sensitive to noise which is undesired amplitude variations which get involuntarily added to a signal. Noise is easily eliminated in an FM system where a constant carrier amplitude is used. Some of the most common applications of FM include FM radio broadcasting, and two-way mobile radio and marine radios.

PHASE MODULATION

The third type of modulation is phase modulation. Remember that phase shift is a time shift between two sine waves of the same frequency. We can use the information signal to shift the phase of the carrier with respect to the carrier reference. The result of this, however, is a signal that looks virtually the same as an FM signal (Figure 63.3). Phase modulation is usually a little easier to implement electronically than FM so most so-called FM systems typically use phase modulation instead. Both FM and PM are often referred as types of angle modulation.

DIGITAL MODULATION

All of the three basic types of modulation, AM, FM, and PM, can also be used to transmit digital or binary data. However, they are usually given different names. Amplitude modulation of a carrier by a binary signal is usually referred to as amplitude shift keying (ASK). This is illustrated in Figure 63.4A. The binary signal simply shifts the carrier amplitude between two specific levels. A special form of ASK is called on–off keying (OOK). This is illustrated in Figure 63.4B. Here the binary signal simply turns the carrier on for a binary 1 and off for a binary 0.

Frequency modulation of a carrier by a binary signal is called frequency shift keying (FSK). Here the binary signal shifts the carrier between two discrete frequencies, see Figure 63.4C.

Phase modulation of a carrier by a binary signal is referred to as phase shift keying (PSK). The term binary PSK (BPSK) is also used. The phase of the carrier is changed as the binary signal switches from 0 to 1 or 1

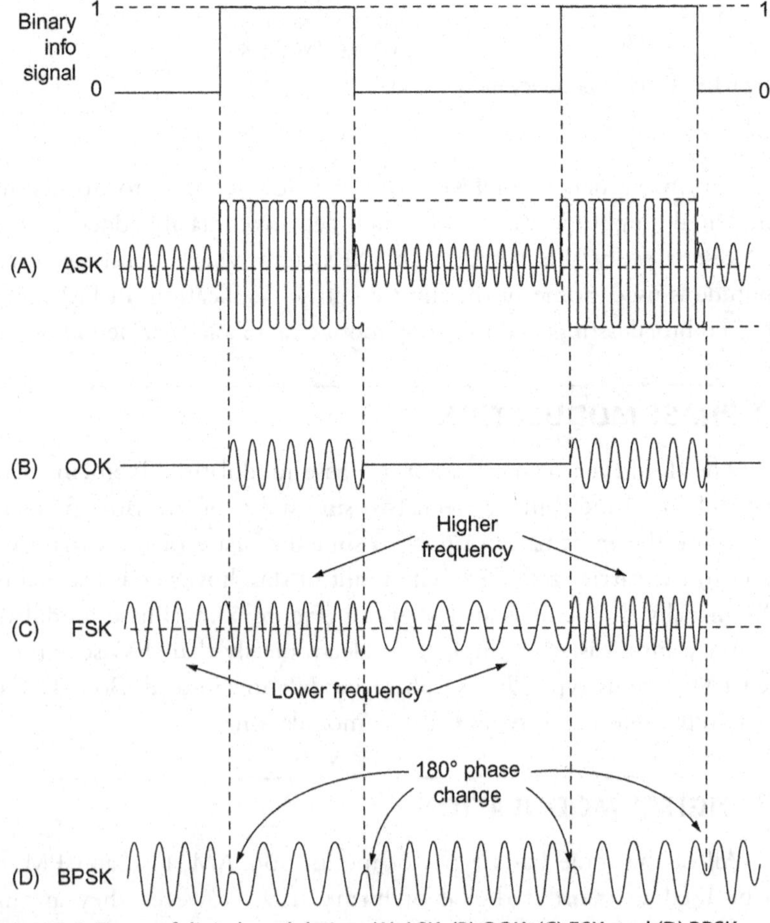

Figure 63.4 Types of digital modulation: (A) ASK, (B) OOK, (C) FSK, and (D) BPSK.

to 0, see Figure 63.4D. Phase shift is 180° which is easily detected at the receiver.

Because binary signals contain many higher frequency harmonics, the resulting signals produced by modulation have an enormous bandwidth. Many schemes have been developed to limit the spectrum of the modulating signal and, as a result, to reduce the amount of bandwidth occupied by a binary modulated signal.

QUADRATURE AMPLITUDE MODULATION

One of the most widely used forms of digital modulation is QAM. QAM is really a composite of both ASK and PSK. The binary signal is converted into a multilevel digital signal which simultaneously modifies both the amplitude and the phase of the signal. QAM is used in digital data transmission in satellites and in modems sending digital data in audio form over the telephone lines or cable TV systems. It is also widely used in the newer digital wireless networks.

A widely used form of QAM is quadrature phase shift keying (QPSK). It processes or modulates two bits at a time. Each bit of a two bit sequence phase modulates the carrier. One carrier is phase shifted 90° with respect to the other. The two PSK signals are then added together to form the output. Each two bit sequence produces one output or symbol.

QPSK is usually illustrated with a constellation diagram as shown in Figure 63.5. The length of the phasor represents the amplitude of the signal while its position represents its phase. Each phase amplitude symbol represents two bits. The overall effect of this is to allow higher data rates in the same bandwidth.

QAM can be implemented to process multiple bits at a time. For example, 16QAM has 16 phase amplitude symbols. Each symbol represents four bits at a time. The constellation diagram is shown in Figure 63.6. Again the result is the ability to transmit higher speeds within the same bandwidth. This benefit is known as spectral efficiency. Spectral efficiency is stated in terms of bits per hertz rate per hertz of bandwidth or b/Hz/Hz.

Keep in mind that the symbol rate is different from the data rate. The symbol rate is also known as the baud rate. The baud rate is lower than the data rate.

There are other higher levels of QAM used in many wired and wireless communications systems. 64QAM and 256QAM are popular but versions of 1024QAM and 4096QAM are sometimes employed. While such

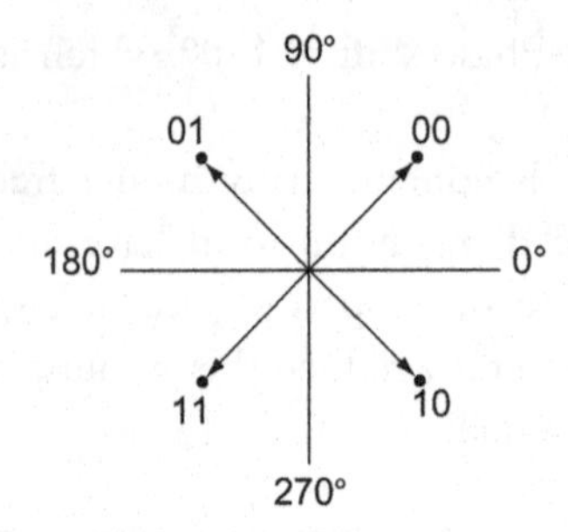

Figure 63.5 The QPSK constellation diagram.

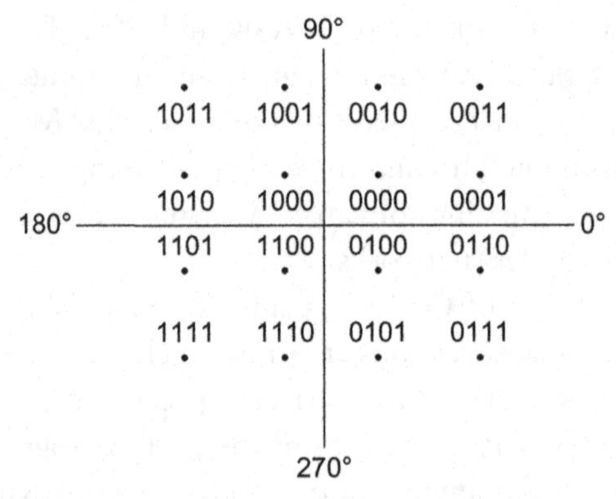

Figure 63.6 16QAM constellation diagram.

systems are very spectrally efficient and permit higher data rates, the different symbols are harder to detect especially in a noisy environment leading to bit errors.

ORTHOGONAL FREQUENCY DIVISION MULTIPLEXING

Orthogonal Frequency Division Multiplexing (OFDM) is another widely used modulation method used to achieve high data rates and spectral efficiency. It is known as a multicarrier modulation method as many carriers are used instead of just one. OFDM takes the serial data to be transmitted and divides it up into many slower serial streams each of which is modulated onto one of multiple subcarriers. There may be as few as 40 subcarriers or many thousands. The subcarriers are spaced from one another by an amount that makes them orthogonal to one another. This

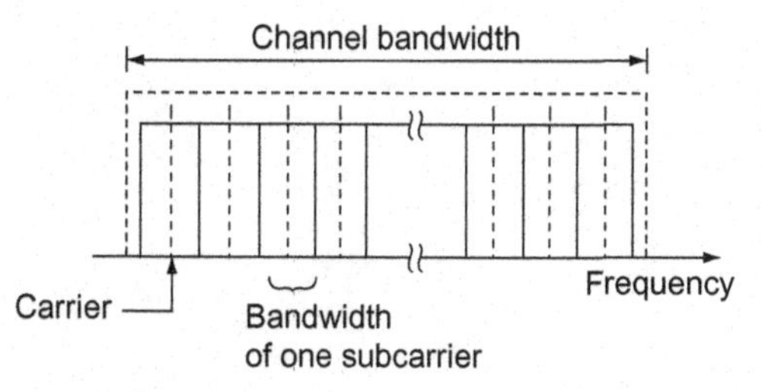

Figure 63.7 OFDM spectrum.

means that despite their close adjacent spacing from one another they will not interfere with one another.

Figure 63.7 shows the spectrum of an OFDM signal. The entire set of carriers is transmitted simultaneously within the assigned bandwidth. The modulation on each carrier is usually BPSK, QPSK, or some form of QAM. This makes OFDM one of the most spectrally efficient of all modulation methods. Furthermore, it is highly resistant to noise and various propagation effects that afflict the transmission channel. OFDM is found in most modern wired and wireless systems today.

BROADBAND MEDIA

Broadband signals are carried over a wide range of wired media. Some use the unshielded twisted pair of the telephone system. Others use cable TV wiring, a combination of fiber optic cable, and coax cable. Some systems use coax cable only.

One unexpected type of medium is the AC power line. The power mains from utility generators to home wiring can carry broadband signals. Known as power line communications (PLC), the power lines are a ready and available medium that requires no new wiring. The broadband signals are simply superimposed on the 50 or 60 Hz sine waves. The main disadvantages are high noise and severe attenuation over distance. Yet, technologies have been developed to overcome these problems. Multiple PLC technologies have been developed to make use of this convenient medium that exists almost everywhere.

CHAPTER SIXTY-FOUR

Data over Cable Service Interface Specifications (DOCSIS)

APPLICATIONS

Cable TV systems delivering and managing video, IP telephony, and Internet services.

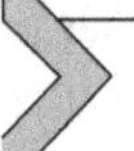

SOURCE

CableLabs.

NATIONAL OR INTERNATIONAL STANDARD

- CableLabs, DOCSIS 3.1 latest version.
- International Telecommunications Union (ITU) Recommendations J.112, J.122, J.222.

KEY FEATURES

- Uses the spectrum on the cable up to about 1 GHz. See Figure 64.1.
- Uses the spectrum as originally designed for TV stations with multiple adjacent 6 MHz wide channels in the United States (8 MHz in Europe).
- Provides for channel bonding that allows the use of two or more channels simultaneously to achieve higher data rates. Typical systems bond 4, 8, 16, or 24 channels to reach up to 10 Gb/s.
- Latest version supports IPv6 addressing.

Handbook of Serial Communications Interfaces.
Doi: http://dx.doi.org/10.1016/B978-0-12-800629-0.00064-4

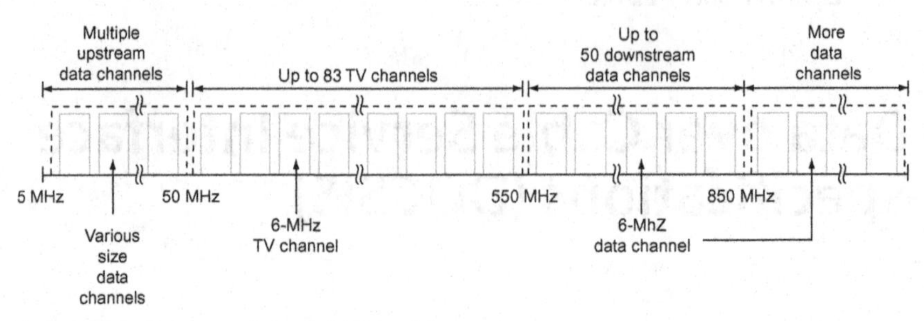

Figure 64.1 Concept of the DOCSIS cable spectrum.

MODULATION METHODS

Varies widely with the system. DOCSIS supports an extensive set of options.

- Downstream TV channels typically use 64QAM in a 6 MHz channel to achieve a 31.2 Mb/s data rate for compressed high-definition TV. 256QAM is an option.
- Downstream data channels use 64QAM or 256QAM with an option to 4096QAM. OFDM is defined for the future.
- Upstream uses QPSK or up to 64QAM using 200 kHz, 3.2 MHz, or 6.4 MHz wide channels.

DATA RATE

Varies widely with the system. DOCSIS supports an extensive set of options.

- 42.88 Mb/s per 6 MHz channel typical maximum.
- Internet services are usually available for up to 25, 50, 100, and 300 Mb/s for different rates depending upon the cable company offerings.

CABLE MEDIUM

Uses a hybrid fiber and coax cable (HFC) system with fiber from the cable modem termination system (CMTS) or cable head end to the neighborhood optical distribution nodes then RG-6/U 75-Ω coax cable to 500–2000 homes or offices. Refer to Figure 64.2.

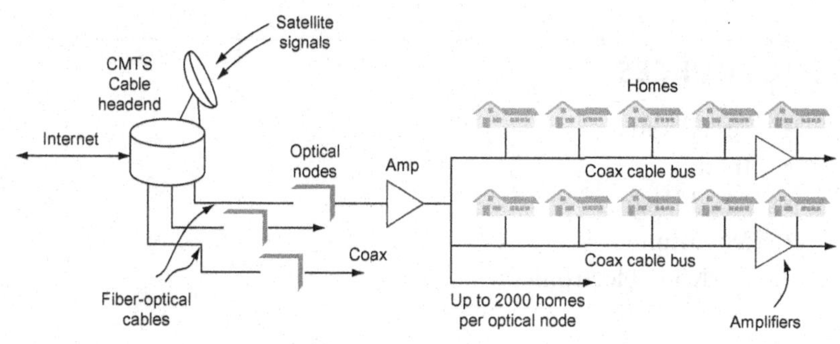

Figure 64.2 Typical cable TV network organization.

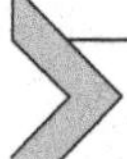

RANGE

Fifty miles absolute maximum, 15 miles maximum typical, depending upon the system.

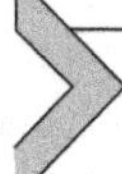

NETWORK CONFIGURATION

- Point-to-point from cable head end to optical node.
- Multidrop bus from optical nodes to homes or offices with amplification added as needed.

PROTOCOL

- Downstream uses frequency division multiple access (FDMA) and time division multiplexing.
- Upstream uses time division multiple access (TDMA).
- Downstream data is packet-based using 188 byte Motion Picture Experts Group (MPEG) packet with 4 bytes of header and 184 bytes of data. Reed-Solomon forward error correction (FEC) is used with the packet.
- Upstream data uses ATM packets with 53 bytes, 5 bytes of header and 48 bytes of data.
- Complex multilayer protocol stack.
- IP addressing is used.
- 56-bit DES and 128-bit AES encryption are used for security.

IC SOURCES

- Broadcom
- Intel
- MaxLinear
- Microtune
- ST Microelectronics

CHAPTER SIXTY-FIVE

Digital Subscriber Line (DSL)

APPLICATIONS

Broadband Internet access, VoIP.

SOURCE

Telecommunications industry, Telecordia Technologies patent 1988.

NATIONAL OR INTERNATIONAL STANDARD

- American National Standards Institute (ANSI): T1-413.
- International Telecommunications Union (ITU): G.992, G.993, G.9700/9701.

KEY FEATURES

- Uses the existing public switched telephone network (PSTN) twisted pair wiring.
- Supports data rates up to 1 Gb/s.
- Most widely used Internet access method worldwide but not in the United States.
- Multiple variations support a wide range of data speeds and ranges (ADSL, ADSL2, VDSL, etc.)
- Uses spectrum of the wiring above the voice band from 300 Hz to 4 kHz in some cases up to 106 MHz. See Figure 65.1.
- DSL service is typically asymmetrical, that is, higher download (DL) speed than upload (UL) speed.

MODULATION METHOD

Discrete multitone (DMT) a form of OFDM (4312.5 Hz subchannels) with BPSK, QPSK, and QAM.

Handbook of Serial Communications Interfaces.
Doi: http://dx.doi.org/10.1016/B978-0-12-800629-0.00065-6

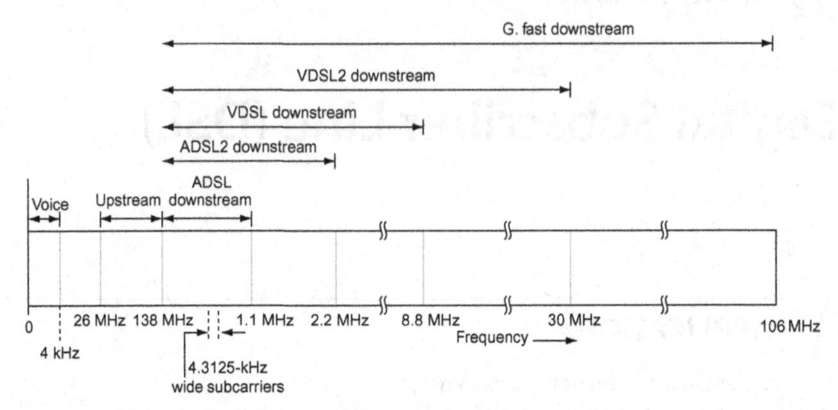

Figure 65.1 The spectrum of the unshielded twisted pair cable showing the subcarriers and the upstream and downstream allocations for ADSL, ADSL2, VDSL, VDSL2, and G.fast.

DATA RATE

Depends upon length of line from central office or Digital Subscriber Line Access Multiplexer (DSLAM) to subscriber. Maximum DL/UL rates:

- Asymmetrical DSL: 8/1 Mb/s
- ADSL2: 12/3.5 Mb/s
- ADSL2+: 24/3.5 Mb/s
- Very high bit rate DSL (VDSL): 52/16 Mb/s
- VDSL2: 200 Mb/s (sum of asymmetrical DL and UL rates)
- G.fast: 1 Gb/s (sum of asymmetrical DL and UL rates)

CABLE MEDIUM

Unshielded twisted pair telephone cable.

RANGE

Maximum typical cable length for highest data speed, from DSLAM to subscriber.

- ADSL 1.2 km
- ADSL2 8000 ft
- ADSL2+ 4000 ft
- VDSL 1000 ft
- VDSL2 500 ft
- G.fast 250–400 ft

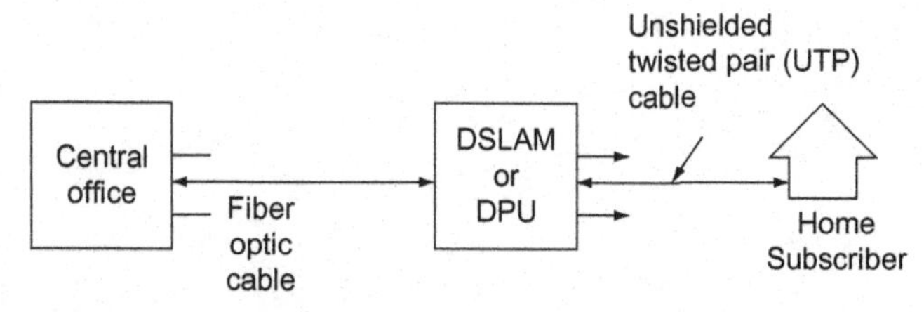

Figure 65.2 DSL network connections.

5 bytes	48 bytes
Header	Data

Figure 65.3 ATM packet.

NETWORK CONFIGURATION

See Figure 65.2.

- Point-to-point from central office to subscriber or
- Point-to-point from DSLAM to subscriber.
- DSLAM in a neighborhood to central office connection is usually fiber optic cable.

PROTOCOL

Asynchronous transfer mode (ATM). ATM packet is shown in Figure 65.3.

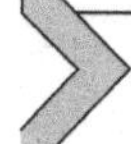

IC SOURCES

- Infineon
- Intersil
- LSI Logic
- Maxim Integrated
- NEC
- Pulse
- ST Microelectrornics

CHAPTER SIXTY-SIX

G3-PLC

APPLICATIONS

- Power Line Communications (PLC) on AC power mains.
- Smart grid monitoring and control.
- Utility metering.
- Home networking.
- Lighting.
- Alternative energy (solar, wind) monitoring and control.

SOURCE

Original development with Electricite Reseau Distribution France (ERDF), Maxim Integrated, and Sagemcom.

NATIONAL OR INTERNATIONAL STANDARD

- International Telecommunications Union (ITU): G.9903.
- Institute of Electrical and Electronic Engineers (IEEE): 1901.2.

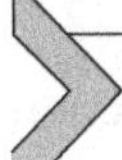

KEY FEATURES

- Designed for low-speed digital data communications for monitoring and control operations in industry or home.
- Highly resistant to noise.
- Designed to pass through low-voltage/medium-voltage (LV/MV) distribution transformers.
- Highly robust and reliable.
- Similar to IEEE 1901.2 PLC standard.

Handbook of Serial Communications Interfaces.
Doi: http://dx.doi.org/10.1016/B978-0-12-800629-0.00066-8

MODULATION METHOD

OFDM with DBPSK, DQPSK, or D8PSK.

DATA RATE

- Depends upon bandwidth used.
- CENELEC A-band (3–95 kHz) 46 kb/s.
- FCC PLC band (10–490 kHz) 250–300 kb/s.

CABLE MEDIUM

LV and MV AC power distribution lines.

RANGE

Typically several hundred meters or less. Two to six miles maximum.

NETWORK CONFIGURATION

- Multidrop bus.
- Support for mesh networking.

PROTOCOL

- Dual forward error correction (FEC) schemes: Reed-Solomon and Vertibi.
- Uses carrier sense multiple access with collision detection (CSMA/CA).
- Uses IETF 6LoWPAN standard to support IPv6 addressing.
- AES-128 cryptography.

IC SOURCES

- Atmel
- Freescale
- Maximum Integrated
- ST Microelectronics
- Texas Instruments

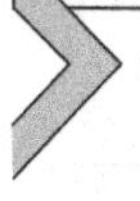

CHAPTER SIXTY-SEVEN

G.hn

APPLICATIONS

Any and all home networking uses.

- Video transport
- Internet access
- Home monitoring and control

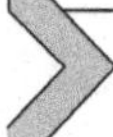

SOURCE

International Telecommunications Union (ITU).

NATIONAL OR INTERNATIONAL STANDARD

- International Telecommunications Union: G.9960, G.9961.
- HomeGrid Forum.

KEY FEATURES

- Can use any existing home wiring, cable TV, telephone, or AC power.
- Capable of data rates to 1 Gb/s.
- Multiple input multiple output (MIMO) option available for AC power lines.

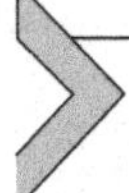

MODULATION METHOD

OFDM with QAM.

Handbook of Serial Communications Interfaces.
Doi: http://dx.doi.org/10.1016/B978-0-12-800629-0.00067-X

DATA RATE

Varies with medium. Data rates to 1 Gb/s.

CABLE MEDIUM

- Installed cable TV coax
- Installed unshielded twisted pair telephone cable
- AC power line

RANGE

Throughout a home.

NETWORK CONFIGURATION

Bus.

PROTOCOL

- Packet-based.
- Time division multiple access (TDMA) for media access with contention-based time slots.
- Low-density parity check (LDPC) FEC.
- Supports ARQ retransmission.
- Supports node to node relays.
- Security via AES-128 encryption.
- Supports IPv4 and IPv6.

IC SOURCES

- Lantiq
- Marvell
- Metanoia
- Sigma Designs

CHAPTER SIXTY-EIGHT

HomePlug (HP)

APPLICATIONS

Any and all home networking uses.

- Video transport
- Internet access
- Monitoring and control
- Electrical smart grid

SOURCE

HomePlug Alliance.

NATIONAL OR INTERNATIONAL STANDARD

- HomePlug Alliance.
- Institute of Electrical and Electronic Engineers (IEEE): 1901 and 1905.1.

KEY FEATURES

- Dominant method and standard of home power line communications (PLC).
- Different versions for different applications: HP-AV, HP-AV2, HP-Green PHY.

MODULATION METHOD

OFDM with QAM.

Handbook of Serial Communications Interfaces.
Doi: http://dx.doi.org/10.1016/B978-0-12-800629-0.00068-1

DATA RATE

- HP-AV 200 Mb/s peak
- HP-AV2 500 Mb/s to 1 Gb/s peak
- HP-GreenPHY 10 Mb/s peak

CABLE MEDIUM

AC power lines.

RANGE

Throughout a home.

NETWORK CONFIGURATION

Bus.

PROTOCOL

- Proprietary
- Packet-based
- Turbo code FEC
- AES-128 encryption

IC SOURCES

- Broadcom
- Qualcomm Atheros
- Sigma Designs
- SPiDCOM

CHAPTER SIXTY-NINE

Multimedia over Cable Alliance (MoCA)

APPLICATIONS

- Interconnecting HDTV sets, cable or satellite set top boxes, and DVRs.
- Other home networking uses.
- Wireless network extension.

SOURCE

Multimedia over Cable Alliance.

NATIONAL OR INTERNATIONAL STANDARD

Multimedia over Cable Alliance, latest version 2.0.

KEY FEATURES

- Uses existing installed home cable TV coax cable for networking.
- Very high-speed capability for video transfer or Internet access.
- Packet error rate less than 10^{-8}.

MODULATION METHOD

OFDM, over 500–1650 MHz of cable spectrum.

DATA RATE

Version 2.0.

Basic: Gross rate 700 Mb/s, 400 Mb/s net throughput.

Handbook of Serial Communications Interfaces.
Doi: http://dx.doi.org/10.1016/B978-0-12-800629-0.00069-3

Enhanced: Gross rate 1.4 Gb/s, 700 Mb/s net throughput.
Capable of 1 Gb/s net rate with channel bonding.

CABLE MEDIUM

Existing cable TV coax, usually RG-6/U 75 Ω or equivalent.

RANGE

Throughout the home.

NETWORK CONFIGURATION

Bus.

PROTOCOL

Proprietary.

IC SOURCES

- Broadcom
- Entropic
- ViXS

CHAPTER SEVENTY

PoweRline Intelligent Metering Evolution (PRIME)

APPLICATIONS

- Smart grid monitoring and control.
- Advanced metering infrastructure (AMI) utility meter monitoring.
- Asset monitoring.

SOURCE

PRIME Alliance.

NATIONAL OR INTERNATIONAL STANDARD

- International Telecommunications Union (ITU) G.9904.
- Institute of Electrical and Electronic Engineers (IEEE) 1901.2.

KEY FEATURES

- Uses in-place electrical power lines.
- Open, nonproprietary, royalty-free standard.
- Latest version is 1.4.

MODULATION METHOD

OFDM using DBPSK, DQPSK, and D8PSK.

DATA RATE

Depends upon spectrum and modulation method.

- 5.4 kb/s using DBPSK in 42–89 kHz CENELEC A-band.

Handbook of Serial Communications Interfaces.
Doi: http://dx.doi.org/10.1016/B978-0-12-800629-0.00070-X

- 128.6 kb/s using D8PSK.
- Up to 1 Mb/s using US FCC band up to 490 kHz.

CABLE MEDIUM

Low-voltage and medium-voltage AC power lines.

RANGE

Up to 1.5 km.

NETWORK CONFIGURATION

Tree topology.

PROTOCOL

- Uses subnetworks with the tree topology.
- Base and Service nodes are defined. Base nodes manage subnetworks.
- Supports two frame types, data and ACK/NACK.
- Uses CSMA/CA for access during contention periods.
- Uses dynamic addressing to reduce overhead.
- Size of CRC depends upon the band used.
- Supports ARQ.
- Uses AES-128 CCM for security.
- Supports IPv4 and IPv6.

IC SOURCES

- Atmel
- HiTrend
- Indra
- Phoenix Systems
- Renesas
- Semitech Semiconductor
- ST Microelectronics
- Texas Instruments

CHAPTER SEVENTY-ONE

X10 Interface

APPLICATIONS

Home automation such as control of lighting and appliances and home security.

SOURCE

Developed by Pico Electronics in 1975.

NATIONAL OR INTERNATIONAL STANDARD

Maintained by Authinx. Previously maintained by the original X10 company X10 WTI (www.X10.com).

KEY FEATURES

- Uses the AC power line as the communications medium (Power Line Communications or PLC).
- Uses digital codes to control multiple devices such as lamps, fans, thermostats, IR sensors, appliances.
- A wireless version is available.

MODULATION METHOD

ASK/OOK.

DATA RATE

120 bps (based on bits occurring at successive zero-crossings at 60 Hz: 8.333 ms).

Handbook of Serial Communications Interfaces.
Doi: http://dx.doi.org/10.1016/B978-0-12-800629-0.00071-1

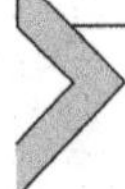

CABLE MEDIUM

Home AC electrical wiring.

RANGE

Complete home coverage. Some "leakage" to adjacent homes may occur.

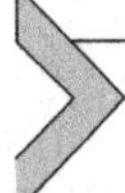

NETWORK CONFIGURATION

- AC house wiring is a bus.
- A master controller or PC manages the control of multiple slaves.
- Digital control codes are transmitted as bursts of 120 kHz sine waves at the zero-crossing points of the 120V 60 Hz power line voltage.
- A typical system consists of a controller that uses pushbuttons to operate up to 16 devices. An electronic switching device that plugs into the AC outlet controls a lamp or other device.

LOGIC LEVELS

Binary 1: 1 ms burst of 120 kHz sine wave at 60 Hz zero-crossing points. Binary 0: Absence of 1 ms burst of 120 kHz sine wave at 60 Hz zero-crossing points.

PROTOCOL

- The protocol frame or packet consists of a start code, house or letter code, followed by a function code.
- The start code is 1110.
- The house or letter code is part of a 4-bit address, 0000-1111.
- The function code is a 5-bit number designating either an address extension of the controlled unit or a command. The last bit determines the function, 0 for unit address or 1 for a command.

- Each packet or frame is transmitted twice to ensure reliable communications under noisy conditions.
- Packets are separated by a 6-bit code 000000.

IC SOURCES

No special X10 ICs are available.

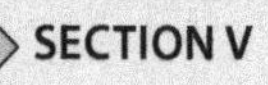

Wireless Interfaces

CHAPTER SEVENTY-TWO

Wireless Interfaces

When a cable is not available or just impractical, a wireless link can serve as a cable replacement for digital data transmission. A wireless link is a free space path between the source transmitter and the destination receiver. The data is serial and modulates a carrier that is converted into an electromagnetic or radio wave by the antenna. There are multiple standards for wireless interfaces, most created for specific applications. This section of the book summarizes the most popular wireless technologies focusing on the factors that are most important in selecting or otherwise working with wireless.

FREQUENCY

Frequency of operation defines where in the electromagnetic spectrum the signal is located. The radio spectrum is very wide from roughly 100kHz to 100GHz. However, most wireless links occupy the very high frequency (VHF) range from 30 to 300MHz, the ultrahigh frequency (UHF) range from 300MHz to 3GHz, or the microwave frequencies above 1GHz.

There is both licensed and unlicensed spectrum. Licensed spectrum is granted primarily to cellular companies by the Federal Communications Commission (FCC) in the United States. Unlicensed spectrum has been allocated for use by anyone. The FCC's rules and regulations designated Parts 15 and 18 provide the guidelines for unlicensed spectrum. Other countries have similar regulatory agencies and relevant rules and regulations.

RANGE

Most wireless interfaces are designed for short-range applications. The term "short-range" means distances from a few centimeters to several miles. While wireless data links can achieve distances of any size, most cover distances of several hundred meters or less. Typical interfaces

Handbook of Serial Communications Interfaces.
Doi: http://dx.doi.org/10.1016/B978-0-12-800629-0.00072-3

are good for 10–100 m maximum. Cellular interfaces are good for several kilometers.

POWER

The amount of transmitter power determines the range and reliability of a wireless link. The more power, the better. However, most regulations stipulate low power, mostly below 1 W or 30 dBm. (Decibels referenced to 1 mW.) Most short-range links use even less power from 1 mW (0 dBm) to 100 mW (20 dBm). Cellular base stations use more power up to about 40 W. Low power is necessary not only to meet regulations but also to limit power consumption in portable devices and to minimize interference to other users.

RECEIVER SENSITIVITY

Another key factor in achieving maximum range is receiver sensitivity. This is a measure of the lowest signal level a receiver is capable of recognizing and recovering reliably. Receiver sensitivity is usually expressed in terms of – dBm. Such – dBm figures represent nanowatts to microwatts of signal level. Typical receiver sensitivities run from roughly −80 to −150 dBm. The higher the number, the better the sensitivity.

PROPAGATION

Propagation refers to the transit of the radio signal from the transmitter antenna to receiver antenna. The path of the signal is critical to achieving maximum range. VHF, UHF, and microwave signal propagation is known as line of sight (LOS). The transmit antenna must "see" the receive antenna. There should be an unobstructed path for reliable signal connectivity. While radio waves do pass through some objects the signal is greatly attenuated. Trees are a good example. The signal will also be reflected by metallic objects. Metal objects such as buildings, water towers, cars, or planes can actually block a signal completely. As a general rule, the LOS path should be clear. It is always best to mount any antennas as high as possible to improve the LOS path.

Wavelength is also a factor in propagation. Longer wavelength signals travel farther than shorter wavelength signals. Remember, wavelength (λ) is related to frequency by the expression:

$$\lambda = 300/f_{MHz}$$

For example, a signal at 2.4 GHz (2400 MHz) has a wavelength of:

$$\lambda = 300/2400 = 0.125 \text{ meters or } 12.5 \text{ cm}$$

The higher the frequency of the signal, the shorter its range for a given power and receiver sensitivity.

This relationship is summed up by the so-called Friis formula:

$$P_r = P_t G_t G_r \lambda^2 / 16\pi^2 d^2$$

P_r is the power received, P_t is the transmitted power in watts, G_t and G_r are the transmit and receive antenna gains (usually 1), λ is the signal wavelength, and d is the distance from transmitter to receiver in meters. As you can see by interpreting the formula for distance, the longer the wavelength the greater the range. A 900 MHz signal will travel farther than a 2400 MHz (2.4 GHz) signal all other factors being the same.

NOISE

A major factor in all wireless transmissions is noise. Random noise will interfere with the signal. If it does not mask the signal completely, it will otherwise cause bit errors, shorten the transmission range, and lower effective data rate.

Noise comes from many sources including electrical interference from power lines, electrical equipment, auto ignitions, and other nearby radio sources. Thermal noise from resistors and other electronic components is especially troublesome when low signal levels are common. A high signal to noise ratio (SNR or S/N) is important to ensure reliable connectivity. An SNR of 30 dB or more is recommended for best transmission.

MODULATION

The short-range wireless interfaces use the same modulation techniques described in the Broadband chapter. ASK, FSK, and BPSK are used as are variations of each. QAM and OFDM are used in some of the more sophisticated standards for higher data rates. One other modulation method not mentioned earlier is spread spectrum that is used in some short-range standards.

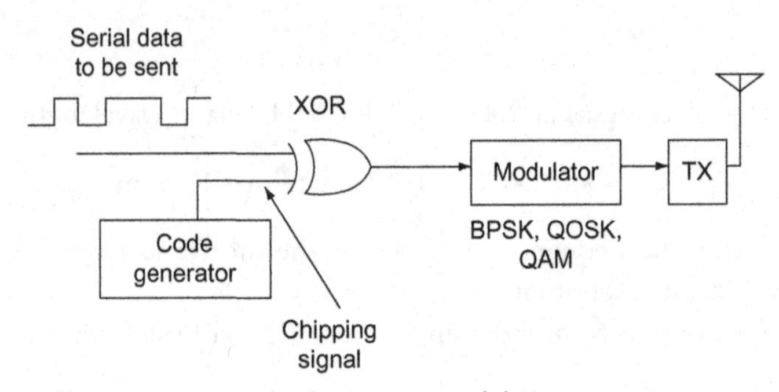

Figure 72.1 Direct sequence spread spectrum modulation.

Spread spectrum (SS) is a technology that allows multiple signals to share the same spectrum space. It makes more efficient use of existing spectrum. There are two types of SS, direct sequence spread spectrum (DSSS) and frequency hopping spread spectrum (FHSS). DSSS is also known as code division multiple access (CDMA). What it does is to use a special coding technique unique to each user. The basic technique is illustrated in Figure 72.1. The serial data to be transmitted is exclusive ORed (XOR) with a higher data rate coded word called the chipping signal. The chipping signal comes from a pseudorandom code generator. The resulting XOR output signal is then used to modulate the carrier usually with BPSK, QPSK, or some form of QAM. The random code spreads the modulation bandwidth over a wide frequency. An example is the 3G CDMA cellular system that uses a 3.84 MHz chipping signal that spreads the signal over a 5 MHz wide channel. To recover the signal, the receiver must know the code so that it can recognize the signal over the assigned bandwidth.

Another form of SS is FHSS. In this form of modulation, the carrier frequency is rapidly changed from one random frequency to another within an assigned bandwidth. A common type of modulation is FSK. The FSK data signal is effectively chopped up and transmitted in segments at a higher rate thereby spreading the signal. To recover the signal, the receiver must have the same hop sequence.

Multiple CDMA or FHSS signals may occupy the same bandwidth at the same time without interfering with one another. Each user has its own unique code that the receiver can identify and reconstruct. Otherwise, a receiver will simply interpret the multiple mixed signals as random noise.

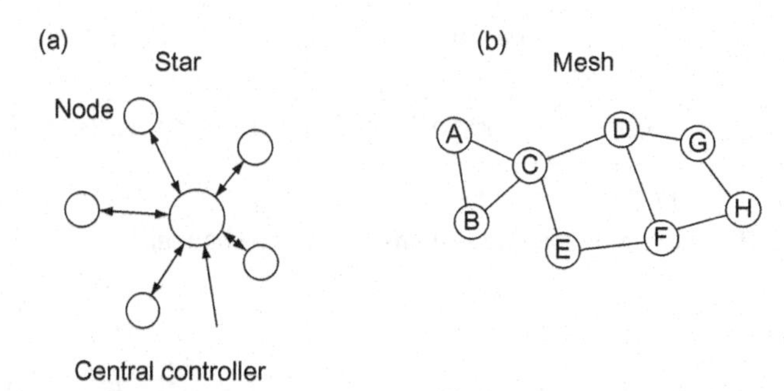

Figure 72.2 Common wireless network topologies: (a) STAR and (b) MESH.

NETWORKING

Wireless signals may also form networks. Many wireless links are simply point-to-point connections between two users. Multiple wireless stations can also talk to one another or a master control point by various networking connections. The two most common networking technologies are star and mesh.

The star connection shown in Figure 72.2(a) uses a central control station. Each wireless node can communicate directly with the central station. Also one node may speak to another node but only by going through the central station.

The mesh network provides multiple links between individual nodes as shown in Figure 72.2(b). One node may speak only to those nodes nearby. However, any node can speak to any other node or the master control node by finding a path through the available connections. This mesh feature extends the range a node can have and also provides redundant paths. If one node fails to relay a signal, a new path can usually be found through the mesh. Mesh networks improve the overall network reliability and are said to be self-healing if one node fails.

ACCESS

When multiple signals want to access the same frequency or bandwidth, some form of access control is used. A common method is carrier sense multiple access with collision avoidance (CSMA/CA). Each node's receiver listens to the channel for a signal. If the channel is open, the node

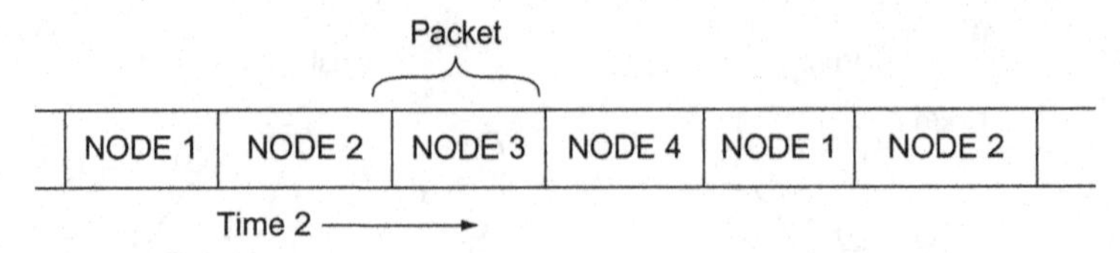

Figure 72.3 TDMA lets multiple nodes share a common channel.

can transmit. If the channel is occupied, the node backs off for a short random time then tries again until access is finally attained.

Another form is channel access is time division multiple access (TDMA). Each node in the network shares the same frequency and bandwidth but is given a dedicated time slot when it can transmit. Figure 72.3 shows a four-node network time transmission. Each packet gets its own transmit time period. All nodes receive all time slots. The periods repeat.

PROTOCOL

Wireless interfaces also have specific protocols for transmitting, receiving, and networking access. Data is always transmitted in packets with a unique frame format. Like wired protocols, the frame uses synchronizing bit sequences, addresses, and data fields. Most also use a frame check sequence or CRC for error detection. Some also include forward error correction (FEC) to improve link reliability in the presence of noise. Many standards include security in the form of cryptography. One popular method is the Advanced Encryption Standard AES-128 where a unique 128-bit code key is used in the encryption process.

APPLICATIONS

The number of wireless interface applications is vast. In this book, only short-range data interfaces are covered. Most involve monitoring and/or control functions while others are used for creating a link to a network or the Internet. Each interface covered here lists the most common applications which are for the most part self-explanatory. Both home/personal and industrial applications are covered. Two special cases require some further explanation.

Machine-to-machine (M2M) applications are those uses in business and industry for monitoring and control operations. Some examples are vending machine monitoring, fleet car and truck vehicle monitoring, remote facility monitoring, video surveillance, and industrial networking. A high percentage of M2M applications use a standard cellular telephone data connection. M2M users have contracts separate and different from voice, texting, or email accounts. Items to be monitored or controlled incorporate a cellphone radio module for data only, usually low speed. Some M2M applications may use other wireless options like Wi-Fi.

Another broad category of application is a movement known as the Internet of Things (IoT). This is an effort to connect every practical item to the Internet for monitoring or control. Examples are remote control of a home thermostat, monitoring of home appliances, lighting control, opening a door lock remotely, medical or fitness monitoring, or video surveillance. It is estimated there are already billions of connected devices. One projection estimates 50 billion Internet-linked devices by 2020. Most devices use a wireless interface.

CHAPTER SEVENTY-THREE

802.15.4

APPLICATIONS

- Wireless personal area networks.
- Industrial monitoring and control.
- Home area networks.
- Internet of Things (IoT) connectivity.
- Smart grid, metering.
- Wireless remote control of consumer electronics (RF4CE).

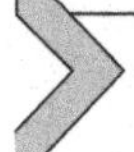

SOURCE

Institute of Electrical and Electronic Engineers (IEEE).

NATIONAL OR INTERNATIONAL STANDARD

IEEE 802.15.4a/b base version.

- 802.15.4c (China)
- 802.15.4d (Japan)
- 802.15.4e (Industrial)
- 802.15.4f (RFID)
- 802.15.4g (Smart Utility Networks – SUN)

KEY FEATURES

- Short range, low speed.
- Low power consumption.
- License-free spectrum.
- Good security.
- Used as the basis for advanced wireless standards (ZigBee, WirelessHART, ISA 100a, etc.)

Handbook of Serial Communications Interfaces.
Doi: http://dx.doi.org/10.1016/B978-0-12-800629-0.00073-5

- Internet Engineering Task Force (IETF) standard 6LoWPAN encapsulation and header compression software is available to handle IPv6 addresses in nodes.

FREQUENCY OF OPERATION

Primary bands:

- 902–928 MHz, Part 15 US unlicensed frequency spectrum.
- 2.4–2.4835 GHz, Part 15 US unlicensed frequency spectrum.
- 868–868.6 MHz, Europe.

MODULATION METHOD

Direct sequence spread spectrum (DSSS) with Differential BPSK or Offset-QPSK.

DATA RATE

- 20 kb/s (868 MHz band)
- 40 kb/s (902–928 MHz band)
- 250 kb/s (2.4 GHz band)

POWER LEVEL

Typically 1 mW (0 dBm).

RANGE

Depends on power level and environment. Typically 10 m up to 100 m under the right conditions.

NETWORK CONFIGURATION

- Point-to-point.
- Star topology.

PROTOCOL

- PHY and MAC layers.
- Packet-based. See frame format in Figure 73.1.

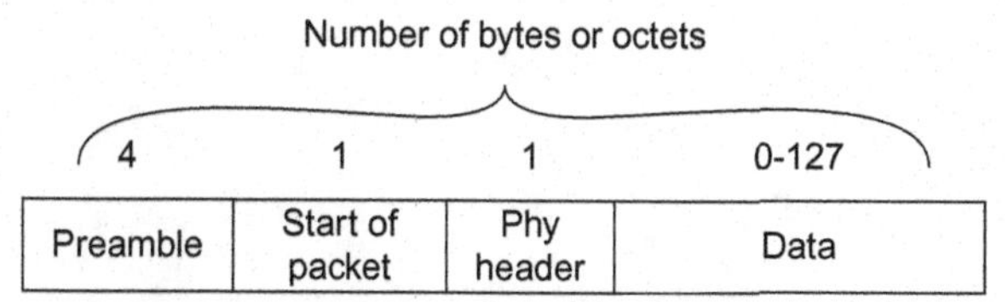

Figure 73.1 Typical IEEE 802.15.4 data packet.

- CSMA/CA access mode.
- 2^{16} addresses for nodes.
- AES-128 encryption for security.

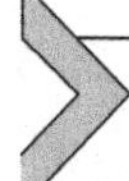

IC SOURCES

- Atmel
- Freescale/NXP
- GreenPeak
- Marvell
- Microchip Technology
- NXP
- Silicon Labs
- Texas Instruments

CHAPTER SEVENTY-FOUR

Bluetooth (BT)

APPLICATIONS

Short-range cord replacement uses. Common uses include:

- Wireless headsets
- Hands-free phone systems in vehicles
- Wireless speakers
- iBeacon proximity sensing
- Location-based services
- Personal area networks (PANs)
- Medical monitoring
- Human interface devices (HID), keyboards, mice, game controllers
- Wristwatch accessories and fitness bands for cell phones
- Internet of Things (IoT)

SOURCE

Invented by Swedish company Ericsson in 1994.

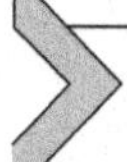

NATIONAL OR INTERNATIONAL STANDARD

Bluetooth Special Interest Group (SIG). Latest version 4.2.
Institute of Electrical and Electronic Engineers (IEEE): 802.15.1.

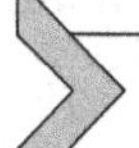

KEY FEATURES

- Most widely used short-range wireless technology (by chip volume).
- Low power consumption.
- Single chip implementation.
- Operates in license-free spectrum.

Handbook of Serial Communications Interfaces.
Doi: http://dx.doi.org/10.1016/B978-0-12-800629-0.00074-7

- Special version Bluetooth Low Energy or Bluetooth Smart for IoT applications.
- Bluetooth SIG defines software profiles for specific applications.

FREQUENCY OF OPERATION

Uses 2.4–2.4835 GHz, Part 15 US unlicensed frequency spectrum.

MODULATION METHOD

Frequency hopping spread spectrum (FHSS) using Gaussian FSK. Other modulation methods for higher data rates include π/4-DQPSK and 8DPSK.

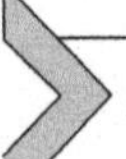

DATA RATE

Depends upon the version and modulation.

- GFSK 1 Mb/s, 2.5 Mb/s option with BT Smart
- π/4-DQPSK 2.1 Mb/s
- 8DPSK 3 Mb/s

POWER LEVEL

- Class 1 100 mW (20 dBm)
- Class 2 2.5 mW (4 dBm), most common
- Class 3 1 mW (0 dBm)
- BT Smart 10 mW (10 dBm)

RANGE

Generally less than 10 m. Up to 100 m maximum with maximum power.

- Class 1 100 m
- Class 2 10 m
- Class 3 1 m
- BT Smart 100 m

NETWORK CONFIGURATION

- Point-to-point
- Piconet with up to seven nodes
- Supports star topology
- Mesh (future addition)

PROTOCOL

- Packet-based.
- Adaptive hopping for interference mitigation.
- Uses IETF 6LoWPAN header compression for IPv6 addressing.
- Forward error correction.
- 24-bit CRC with automatic repeat request (ARQ).
- Encryption and authentication for security. 56–128-bit keys.

IC SOURCES

- Bluegiga
- Broadcom
- Cambridge Silicon Radio
- Nordic Semiconductor
- NXP
- Rabbit
- RFM
- Qualcomm
- ST Microelectronics
- Texas Instruments

CHAPTER SEVENTY-FIVE

Digital Enhanced Cordless Telecommunications (DECT)

APPLICATIONS

- Primary use cordless telephones for home or office.
- PBX service.
- Potential home automation monitoring and control uses.

SOURCE

European Telecommunications Standards Institute (ETSI).

NATIONAL OR INTERNATIONAL STANDARD

ETSI.

KEY FEATURES

- Short range.
- Very low power.
- Simple implementation.
- Very low cost.
- US version DECT 6.0.
- Ultra low energy (ULE) version for IoT.

FREQUENCY OF OPERATION

- 1880–1900 MHz (Europe and worldwide)
- 1920–1930 MHz (the United States)
- 1900–1920 MHz and 1910–1930 MHz (Outside Europe)

Handbook of Serial Communications Interfaces.
Doi: http://dx.doi.org/10.1016/B978-0-12-800629-0.00075-9

MODULATION METHOD

- GFSK most common.
- Allowed alternatives: $\pi/2$ DBPSK, $\pi/4$-DQPSK, $\pi/8$-D8PSK.

DATA RATE

- 32 kb/s voice most common.
- Up to 2 Mb/s data with alternative modulation.

POWER LEVEL

Commonly 10 mW, 250 mW maximum (Europe), 4 mW, 100 mW maximum (the United States).

RANGE

- Full home or office coverage for voice, up to 200 m typical.
- Up to 10 km for data with maximum power.

NETWORK CONFIGURATION

Point-to-point.

PROTOCOL

- 10 carriers, 1.728 MHz spacing (Europe), 5 carriers 1.728 MHz spacing (the United States).
- Uses time division duplexing (TDD).
- Uses time division multiple access (TDMA) to access multiple (12) TDD data channels. 24 time slots, 12 up, 12 down.
- Dynamic channel allocation (DCA).
- Standard DECT frame and time slot: 420 bits and 417 μs long.
- Uses ADPCM codec for voice compression. ITU G.726, G.711, G.722, G.729.

IC SOURCES

- Analog Devices
- Atmel
- Dialog Semiconductor
- DSP Group
- Infineon
- M/A-COM
- Maxim Integrated
- NXP
- ST Microelectronics
- Texas Instruments

CHAPTER SEVENTY-SIX

EnOcean

APPLICATIONS

- Building automation.
- Smart home monitoring and control.
- Industrial control.

SOURCE

EnOcean GmbH, a German company spun off from Siemens AG.

NATIONAL OR INTERNATIONAL STANDARD

- ISO/IEC 14543-3-10.
- EnOcean Alliance.

KEY FEATURES

- An energy-harvesting technology that uses mechanical, thermal, solar, or piezoelectric methods to power the devices. No batteries or other power source needed.
- Wireless devices include switches, sensors, and actuators. Light switches and sensors are the most common.
- Software profiles for specific applications.
- Easily connects to wired networks.
- IP licenses available.

FREQUENCY OF OPERATION

315, 868, and 902–928 MHz unlicensed spectrum.

Handbook of Serial Communications Interfaces.
Doi: http://dx.doi.org/10.1016/B978-0-12-800629-0.00076-0

MODULATION METHOD

ASK.

DATA RATE

125 kb/s.

POWER LEVEL

50 μW.

RANGE

Less than 30 m indoors, less than 300 m in free space. Repeaters available for range extension.

NETWORK CONFIGURATION

Point-to-point and mesh.

PROTOCOL

- Four-layer OSI protocol: physical, data link, network, and application layers.
- Very short packet frame (14 bytes) for energy savings (<1 ms).
- Packets sent three times at pseudorandom times with checksum to reduce collisions and improve reliability.
- 32-bit node ID.
- TCP/IP compatible.

IC SOURCES

Complete modules available only from EnOcean.

CHAPTER SEVENTY-SEVEN

ISA100-11a

APPLICATIONS

- Process monitoring and control.
- Asset management.
- Sensor networks.
- Building automation.
- Renewable energy.
- Transportation.
- Industrial Internet of Things.

SOURCE

International Society of Automation (ISA).

NATIONAL OR INTERNATIONAL STANDARD

- IEEE 802.15.4e.
- Wireless Compliance Institute (WCI) of the Automation Standards Compliance Institute (ASCI).
- IEC 62734.

KEY FEATURES

- Short range, low speed.
- Low power consumption.
- License-free spectrum.
- Good security.
- Requires license for sale of commercial products.
- Mesh networking.
- Certification for interoperability.

Handbook of Serial Communications Interfaces.
Doi: http://dx.doi.org/10.1016/B978-0-12-800629-0.00077-2

FREQUENCY OF OPERATION

2.4–2.4835 GHz, Part 15 US unlicensed frequency spectrum.

MODULATION METHOD

Direct sequence spread spectrum (DSSS) with Offset-QPSK.

DATA RATE

250 kb/s.

POWER LEVEL

Typically 1 mW (0 dBm).

RANGE

Depends on power level and environment. Typically 10 m up to 100 m under the right conditions.

NETWORK CONFIGURATION

- Point-to-point.
- Star topology.
- Mesh topology, self-healing, self-sustaining.

PROTOCOL

- PHY and MAC layers of IEEE 802.15.4.
- Uses full seven-layer OSI model protocol.
- Packet-based.
- 2^{16} addresses for nodes.
- CSAM/CA access.
- AES-128 encryption for security.
- Internet Protocol (IP) compatibility using 6LoWPAN.
- Extensible to support FOUNDATION Fieldbus, Profibus, HART protocols.

IC SOURCES

See list under 802.15.4.

CHAPTER SEVENTY-EIGHT

Industrial Scientific Medical (ISM) Wireless

APPLICATIONS

- Virtually all short-range wireless uses.
- Mostly remote monitoring and control.

SOURCE

No one original source.

NATIONAL OR INTERNATIONAL STANDARD

- Multiple standards that use the ISM bands (Wi-Fi, Bluetooth, 802.15.4, etc.).
- Many versions with no formal standards.
- FCC Parts 15 and 18.

KEY FEATURES

- Short range.
- Very low power.
- Simple implementation.
- Very low cost.

FREQUENCY OF OPERATION

US FCC Parts 15 and 18 unlicensed ISM frequencies, most common:

- 6.78 MHz
- 13.56 MHz
- 27.12 MHz
- 40.68 MHz

Handbook of Serial Communications Interfaces.
Doi: http://dx.doi.org/10.1016/B978-0-12-800629-0.00078-4

- 315 MHz
- 433.92 MHz
- 868 MHz (Europe)
- 902–928 MHz

MODULATION METHOD

ASK or FSK.

DATA RATE

Depends upon the technology. Typically less than 100 kb/s.

POWER LEVEL

Commonly 1 mW (0 dBm) to 1 W (30 dBm).

RANGE

Depends upon frequency of operation, power level, and the environment. From a few meters up to several kilometers.

NETWORK CONFIGURATION

Point-to-point.

PROTOCOL

Varies widely with the technology.

IC SOURCES

- Analog Devices
- Freescale/NXP
- Infineon
- Maxim Integrated
- Melexis
- Microchip Technology
- Micrel

- Nordic Semiconductor
- Semtech
- Silicon Laboratories
- ST Microelectronics
- Texas Instruments

CHAPTER SEVENTY-NINE

Near Field Communications (NFC)

APPLICATIONS

- Payments, replace credit cards.
- Advertising posters.
- Facilities access.
- Transportation payment.
- Vending machines.
- Automatic pairing of other wireless interfaces.

SOURCE

Sony, Philips (now NXP), and Nokia are the original sources and formed the NFC Forum.

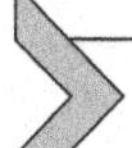

NATIONAL OR INTERNATIONAL STANDARD

- ISO/IEC 18000-3, 18092, 21481, 13157, and 14443
- ECMA-340, ECMA-352
- GSMA
- NFC Forum

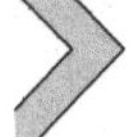

KEY FEATURES

- Very short range for security.
- Uses the near field (magnetic induction) of radio wave propagation.
- Low power.
- Passive, no power operation possible. Chips use RF energy from a reader to create DC power for operation.
- Information storage capability (96–4096 bytes of memory typical).
- Licensing required for use.

Handbook of Serial Communications Interfaces.
Doi: http://dx.doi.org/10.1016/B978-0-12-800629-0.00079-6

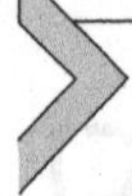

FREQUENCY OF OPERATION

13.56 MHz, unlicensed spectrum.

MODULATION METHOD

ASK either 10% or 100% modulation.

DATA RATE

- 106 kb/s 100% ASK Miller encoding
- 212 kb/s 10% ASK Manchester encoding
- 424 kb/s 10% ASK Manchester encoding
- 848 kb/s possible

POWER LEVEL

1 mW, 0 dBm.

RANGE

Less than 20 cm, 4 cm typical.

NETWORK CONFIGURATION

Point-to-point.

PROTOCOL

- Three types of devices defined: initiator (interrogator), target (tag), reader/writer.
- Initiators can operate full duplex.
- Tags operate half duplex.
- Smart card emulation capability.
- Special security provided.

IC SOURCES

- Infineon
- Inside Secure
- Moversa
- NXP
- ST Microelectronics
- Texas Instruments

CHAPTER EIGHTY

Ultra Wideband (UWB)

APPLICATIONS

- Wireless video transport.
- Computer docking stations.
- Secure short-range military equipment.
- Ground and wall penetrating radar.
- Distance measurement.
- Cable replacement.

SOURCE

Military and university research.

NATIONAL OR INTERNATIONAL STANDARD

- IEEE 802.15.3.
- WiMedia Alliance.

KEY FEATURES

- Short range.
- Very low power.
- Very wide bandwidth (500 MHz or 20% of operating frequency, minimum).
- Multiple types defined (pulse, continuous wave, OFDM).
- Highly secure.
- Versions available for coax cable.

FREQUENCY OF OPERATION

3.1–10.6 GHz unlicensed spectrum.

Handbook of Serial Communications Interfaces.
Doi: http://dx.doi.org/10.1016/B978-0-12-800629-0.00080-2

MODULATION METHOD

Depends upon the technology. OFDM, BPSK continuous wave (CW), and pulse phase shift.

DATA RATE

Depends upon the technology. Typically 480 Mb/s (OFDM), 1.3 Gb/s (CW or pulse).

POWER LEVEL

−41 dBm/MHz.

RANGE

Less than 10 m.

NETWORK CONFIGURATION

Point-to-point.

PROTOCOL

Varies widely with the technology.

IC SOURCES

- Alereon
- Pulse~Link
- Time Domain

CHAPTER EIGHTY-ONE

Wi-Fi

APPLICATIONS

- Local area networks.
- Access points (hot spots).
- Home networks.
- Internet connectivity.
- Public safety operations.
- Backhaul.
- Internet of Things connectivity.
- Cellular data/voice off-load.

SOURCE

Institute of Electrical and Electronic Engineers (IEEE).

NATIONAL OR INTERNATIONAL STANDARD

- IEEE 802.11
- Wi-Fi Alliance

KEY FEATURES

- Short range, high speed.
- License-free spectrum.
- Good security.
- Some versions use MIMO for multipath mitigation and higher data rates.
- Built into most laptop and tablet computers and smartphones.

Handbook of Serial Communications Interfaces.
Doi: http://dx.doi.org/10.1016/B978-0-12-800629-0.00081-4

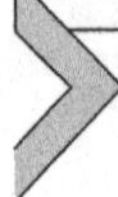

FREQUENCY OF OPERATION

Primary bands:

- 2.4–2.4835 GHz, Part 15 US unlicensed frequency spectrum
- 5.725–5.875 GHz, Part 15 US unlicensed frequency spectrum

Alternate bands:

- 5.18–5.24 GHz
- 5.85–5.925 GHz
- 3.6575–3.6925 GHz
- 470–710 MHz
- 57–64 GHz

MODULATION METHOD

See table.

DATA RATE

See table.

RANGE

See table.

Standard	Frequency (GHz)	Modulation	Maximum Data Rate (Mb/s)	Range (m)
802.11a	5	OFDM	54	50
802.11b	2.4	DSSS	11	100
802.11g	2.4	OFDM	54	100
802.11n	2.4/5	OFDM/MIMO	600	100
802.11ac	5	OFDM/MIMO	1.3 Gb/s	50–100
802.11ad	60	OFDM beamforming	7 Gb/s	10

NETWORK CONFIGURATION

- Point-to-point
- Star topology

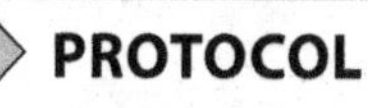

PROTOCOL

IEEE 802.11.

IC SOURCES

- Broadcom
- Infineon
- Intel
- Intersil
- Marvell
- Maxim Integrated
- Qualcomm Atheros
- Texas Instruments

CHAPTER EIGHTY-TWO

WirelessHART

APPLICATIONS

- Process monitoring and control.
- Asset management.
- Sensor networks.
- Building automation.
- Renewable energy.
- Transportation.

SOURCE

HART (Highway Addressable Remote Transducer) Communications Foundation.

NATIONAL OR INTERNATIONAL STANDARD

- IEEE 802.15.4e
- HART Communications Foundation
- IEC 62591

KEY FEATURES

- Wireless version of wired HART technology. See wired HART interface entry.
- Based on the technology of Dust Networks' (Linear Technology).
- Short range, low speed.
- Low power consumption.
- License-free spectrum.
- Good security.
- Requires license for sale of commercial products.
- Mesh networking.
- Certification for interoperability.

Handbook of Serial Communications Interfaces.
Doi: http://dx.doi.org/10.1016/B978-0-12-800629-0.00082-6

FREQUENCY OF OPERATION

2.4–2.4835 GHz, Part 15 US unlicensed frequency spectrum.

MODULATION METHOD

Direct sequence spread spectrum (DSSS) with Offset-QPSK.

DATA RATE

250 kb/s.

POWER LEVEL

Typically 1 mW (0 dBm). 8 or 10 dBm option.

RANGE

Depends on power level and environment. Typically 10 m up to 100 m under the right conditions.

NETWORK CONFIGURATION

- Point-to-point.
- Star topology.
- Mesh topology, self-healing, self-sustaining.

PROTOCOL

- PHY and MAC layers of IEEE 802.15.4.
- Time Synchronized Mesh Protocol (TSMP).
- Uses Time Slotted Channel Hopping (TSCH). TDMA for access.
- Packet-based.
- 2^{16} addresses for nodes.
- AES-128 encryption for security.

IC SOURCES

Linear Technology (Dust Networks).

CHAPTER EIGHTY-THREE

ZigBee

APPLICATIONS

- Wireless personal area networks.
- Industrial monitoring and control.
- Home area networks.
- Internet of Things (IoT) connectivity.
- Smart grid, metering.
- Wireless remote control.
- Lighting control.

SOURCE

ZigBee Alliance.

NATIONAL OR INTERNATIONAL STANDARD

- IEEE 802.15.4a/b base version.
  - 802.15.4c (China)
  - 802.15.4d (Japan)
  - 802.15.4e (Industrial)
  - 802.15.4f (RFID)
  - 802.15.4g (Smart Utility Networks – SUN)
- ZigBee Alliance.

KEY FEATURES

- Uses IEEE 802.15.4 for PHY and MAC.
- Short range, low speed.
- Low power consumption.
- License-free spectrum.
- Good security.
- Requires license for sale of commercial products.

Handbook of Serial Communications Interfaces.
Doi: http://dx.doi.org/10.1016/B978-0-12-800629-0.00083-8

- Adds mesh networking to 802.15.4.
- Certification for interoperability.
- Multiple predeveloped application profiles of software for specific applications.

FREQUENCY OF OPERATION

Primary bands:

- 902–928 MHz, Part 15 US unlicensed frequency spectrum.
- 2.4–2.4835 GHz, Part 15 US unlicensed frequency spectrum. This is the most commonly implemented worldwide version.
- 868–868.6 MHz, Europe.

MODULATION METHOD

Direct sequence spread spectrum (DSSS) with Differential BPSK or Offset-QPSK.

DATA RATE

- 20 kb/s (868 MHz band).
- 40 kb/s (902–928 MHz band).
- 250 kb/s (2.4 GHz band). This is the most commonly implemented worldwide version.

POWER LEVEL

Typically 1 mW (0 dBm). Minimum 0.5 mW or −3 dBm. 100 mW modules available.

RANGE

Depends on power level and environment. Typically 10 m up to 100 m under the right conditions.

NETWORK CONFIGURATION

- Point-to-point.
- Star topology.
- Mesh topology.

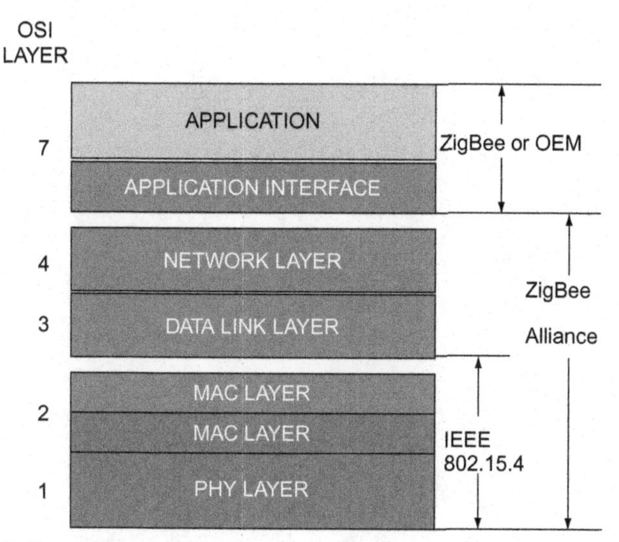

Figure 83.1 ZigBee protocol stack.

PROTOCOL

- PHY and MAC layers of IEEE 802.15.4.
- Adds layers 3 and 7 for networking and application. See Figure 83.1.
- Packet-based. See 802.15.4.
- CSMA/CA access mode.
- 2^{16} addresses for nodes.
- AES-128 encryption for security.

IC SOURCES

- Atmel
- Freescale/NXP
- GreenPeak
- Marvell
- Microchip Technology
- NXP
- Silicon Labs
- Texas Instruments

CHAPTER EIGHTY-FOUR

Z-Wave

APPLICATIONS

- Wireless personal area networks.
- Home area networks.
- Lighting control.
- Internet of Things (IoT) connectivity.

SOURCE

Sigma Designs (Previously Zensys).

NATIONAL OR INTERNATIONAL STANDARD

- Sigma Designs.
- Z-Wave Alliance.

KEY FEATURES

- Short range, low speed.
- Low power consumption.
- License-free spectrum.
- Good security.
- Requires license for sale of commercial products.

FREQUENCY OF OPERATION

908.42 MHz, US FCC Part 15 unlicensed spectrum.

MODULATION METHOD

Gaussian FSK.

Handbook of Serial Communications Interfaces.
Doi: http://dx.doi.org/10.1016/B978-0-12-800629-0.00084-X

DATA RATE

9.6 and 40 kb/s. Up to 100 kb/s possible in some products.

POWER LEVEL

Typically 1 mW (0 dBm).

RANGE

Typically up to 30 m.

NETWORK CONFIGURATION

- Point-to-point.
- Star topology.

PROTOCOL

- Proprietary.
- Up to 232 nodes.
- AES-128 encryption for security.

IC SOURCES

Sigma Designs.

ACRONYM GLOSSARY

6LoWPAN Header compression protocol of the IETF to enable IPv6

ACK Acknowledge. A signal or code that acknowledges receipt of a message or packet

ADM Add-Drop Multiplexer. A hardware device that allows data to be inserted or extracted from an optical network ring

AES Advanced Encryption Standard. A highly secure method developed by the NIST. Uses key sizes of 128, 192, or 256

ANSI American National Standards Institute. A U.S. standards organization

ARQ Automatic repeat request. An error detection mechanism that asks for a message or packet to be sent again

ASCI Automation Standards Compliance Institute of the ISA

ASIC Application-specific integrated circuit. Complex, large-scale chip

ASK Amplitude shift keying. Modulating the amplitude of a carrier with a binary signal by varying its amplitude level

ATM Asynchronous transfer mode. A packet-based protocol

BPSK Binary phase shift keying. A type of digital modulation involving a 180 degree phase shift for each state

CEI .

CENELEC European Committee on Electrotechnical Standards. A key European standards organization for electrical and electronic products

CRC Cyclical redundancy check. A mathematical type of error detection method widely used in electronic communications

CSMA/CA Carrier sense media access with collision detection. Contention-based access method for multiple nodes

DBPSK Differential BPSK. Uses previous bit phase as reference for demodulation

DQPSK Differential QPSK. Uses previous bit phase as reference for demodulation

D8PSK Differential 8PSK. Uses previous bit phase as reference for demodulation

ECMA European Computer Manufacturers Association, now ECMA International

ERDF Electricite Reseau Distribution France

FCC Federal Communications Commission

FTT*x* Fiber to the *x* where *x* means home (H), curb (C), or premises (P), etc.
GSMA GSM Association
HAN Home area network (see PAN)
HV High-voltage AC transmission lines >72 kV
IC Integrated circuit
IEC International Electrotechnical Commission. An international standards organization
IEEE Institute of Electrical and Electronic Engineers. An international professional society and standards organization
IETF Internet Engineering Task Force. Standards organization for networks
IoT Internet of Things
IP Intellectual property, e.g., IC and circuit designs
IR Infrared light
ISA International Society of Automation. An international professional society and standards organization
ISO International Organization for Standardization
ITU International Telecommunications Union. A worldwide standards organization specializing in electronic communications
LAN Local area network. A communications network covering an office, building, or campus
LV Low-voltage AC transmission lines. <1000V
MAC Media access control. Layer 2 of a network protocol
MAN Metropolitan area network. A network used over a local area like a city, neighborhood, or cable TV area
MIMO Multiple input multiple output. The use of multiple transceivers and antennas to achieve higher speeds and mitigation of multipath propagation
MV Medium Voltage AC transmission lines >1000V, <72 kV
NACK Negative acknowledge. Indicates no receipt of a transmission (see ACK)
NIST National Institute of Standards and Technology, the U.S. standards organization
Octet 8-bits, 1 byte
OFDM Orthogonal Frequency Division Multiplexing. A multicarrier modulation method that divides a data stream into many slower data streams and modulates them using some form of PSK or QAM
OIF Optical Internetworking Foundation. A standards organization

OSI Open Systems Interconnection. A seven-layer networking model for communications

PAN Personal area network. Short-range wireless network

PCB Printed circuit board

PHY The physical layer of a network protocol

PON Passive optical network. A LAN or MAN that uses only passive components for transmission

PSK Phase Shift Keying. A form of digital modulation where the binary states are represented by some phase shift angle of the carrier

QAM Quadrature Amplitude Modulation. A combination of both amplitude shift and phase shift modulation used for high speeds in narrow bandwidths

QPSK Quadrature Phase Shift Keying. A form of phase shift modulation where two data streams are mixed with a 90 degree carrier shift between them

SMPTE Society of Motion Picture and Television Engineers. A standards organization

TCP/IP Transmission Control Protocol/Internet Protocol

TDMA Time division multiple access. Synchronized time slots for contentionless access

WAN Wide area network. Large network like the Internet or telephone system

WCI Wireless Compliance Institute. Part of ASCI of ISA

INDEX

Note: Page numbers followed by "*f*" and "*t*" refer to figures and tables, respectively.

X

Z

Printed in the United States
By Bookmasters